A Simple Approach to
Digital Signal Processing

## TOPICS IN DIGITAL SIGNAL PROCESSING

C. S. Burrus and T. W. Parks: *DFT/FFT AND CONVOLUTION ALGORITHMS: THEORY AND IMPLEMENTATION*

John R. Treichler, C. Richard Johnson, Jr., and Michael G. Larimore: *THEORY AND DESIGN OF ADAPTIVE FILTERS*

T. W. Parks and C. S. Burrus: *DIGITAL FILTER DESIGN*

Rulph Chassaing and Darrell W. Horning: *DIGITAL SIGNAL PROCESSING WITH THE TMS320C25*

Rulph Chassaing: *DIGITAL SIGNAL PROCESSING WITH C AND THE TMS320C30*

Craig Marven and Gillian Ewers: *A SIMPLE APPROACH TO DIGITAL SIGNAL PROCESSING*

# A Simple Approach to Digital Signal Processing

Craig Marven and Gillian Ewers

A Wiley-Interscience Publication
JOHN WILEY & SONS, INC.
New York • Chichester • Brisbane • Toronto • Singapore

This text is printed on acid-free paper.



Published simultaneously in Canada.



***Library of Congress Cataloging-in-Publication Data***

Marven, Craig.
A simple approach to digital signal processing / Craig Marven & Gillian Ewers.
p. cm. — (Topics in digital signal processing)
"A Wiley-Interscience publication."
Includes bibliographical references and index.
ISBN 0-471-15243-9 (cloth : alk. paper)
1. Signal processing—Digital techniques. I. Ewers, Gillian.
II. Title. III. Series.
TK5102.9.M38 1996
621.382'2—dc20 96-2518

Printed in the United States of America

10 9 8 7 6 5 4

# Contents

# Acknowledgments

We should like to acknowledge the help of the following in the writing of this book: Gerard Jourdet of Texas Instruments for sponsoring the project, Ross Nimmo formerly of Texas Instruments for his contribution on analog circuitry in Chapter 2, Dr. Andrew Bateman of the University of Bristol for providing many valuable suggestions and helping to make technical corrections to the original proof, Richard Mann of Texas Instruments for helping to correct our grammar and spelling, Edgar Auslander of Texas Instruments for his amendments to the original printing, Kenneth W. Schachter of Texas Instruments for his detailed amendments to the second printing, and most of all Lynda and Tony for putting up with us working through many evenings and weekends.

# Introduction

## THE PURPOSE OF THIS BOOK

This book is intended for anyone who needs to know about digital signal processing (DSP) but hasn't really got the first idea what it's all about.

If you are a "mature" engineer who never studied digital signal processing in college but now need to learn about it, this book is for you. If you are an extremely confused undergraduate, this book is also for you.

We cannot hope to give you all the knowledge you will need to design a high-speed data modem or pass your examinations, but we will provide a readable introduction to this highly technical subject. You will learn about sampling, filters, frequency transforms, data compression and the design decisions you will need to take. In addition, we will point you toward other texts that can extend your knowledge in specific areas.

One of the aims of this book is to give you some basic knowledge so that you can make sense of some of the more advanced books on digital signal processing. So often they start at too high a level. Once you have read this book, you should be ready to tackle them with more confidence.

You will be relieved to learn that there is very little mathematics in this book. We realize that Chapter 5 contains a fair number of equations, but you should see the other books on frequency transforms! We have concentrated on diagrams and explanation as much as possible – you can get all the mathematics you want from the recommendations for further reading.

Finally, if you actually need to design and build a real DSP system, we also have some help for you. A typical system development cycle is explained, along with examples of all the tools necessary to achieve that final working system.

If you really want to know what VSELP means, we also have a handy glossary of DSP-related acronyms.

## WHAT IS DIGITAL SIGNAL PROCESSING?

Our environment is full of signals that we sense, such as sound, temperature and light.

In the case of sound we use our ears to convert it into electrical impulses to our brain. We then analyze properties such as amplitude, frequency, and phase to categorize the sound and determine its direction. We may recognize it as music, speech or a pneumatic drill!

For temperature, nerves in exposed parts of our skin will send signals to the brain. The analysis required is straightforward and may result in us switching on the heating or opening a window.

For light, our eyes focus the image onto our retina, which converts it into an electrical signal to send to the brain. Our brain will analyze the light for color, shapes, intensity, and so on. As in the case of sounds, we will then make an informed decision on recognizing objects, distance, motion and so on.

Of course, our body is equipped with five categories of sensor: hearing, touch, taste, smell and sight and an extremely powerful "computer", our brain. We can make most decisions for ourselves, but there are many cases where we may want a machine to make decisions based on its own sensing of some environmental variable. It is fairly easy to equip the machine with sensors that will convert the variable to an electrical signal. The difficult part comes when we need an electronic computer to act like our own brain.

Our brain works with electrical representations of continuously variable signals such as sound intensity, pressure and so on. These continuously variable signals are known as analog signals. Our brain may be considered an extremely powerful analog computer. Although we can build electronic analog computers, we are much more used to digital computers such as the PC. These are excellent at processing numbers in applications like databases or spreadsheets, but they are

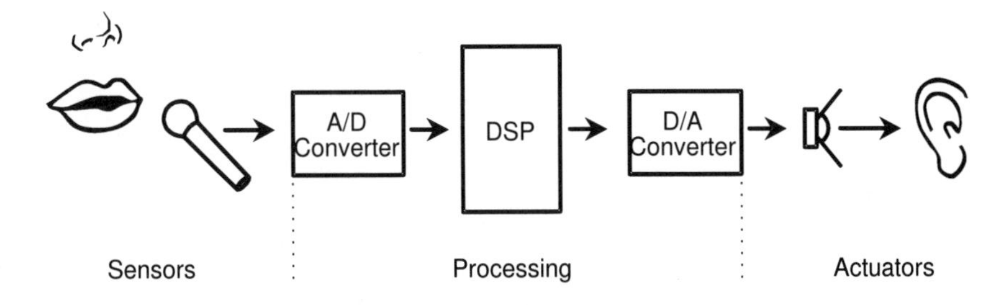


**FIGURE I.1.** The digital signal processing system

not very good at processing the continuously varying signals that make up the world around us.

Our taste buds, nose, ears, eyes and skin convert the signals they receive into electrical energy, which passes along nerves to our brain. We can use electronic sensors to convert pressure, temperature, sound, etc., to electrical signals in the same way, but we still have to change these into numbers to be passed to a digital computer for processing. This conversion process is called analog-to-digital, or A/D, conversion.

The processing that we apply to the signal is carried out by the digital computer and is thus called digital signal processing, DSP for short. In modern DSP systems a single-chip microcomputer specially designed for digital signal processing is usually used. These special microcomputers are known as digital signal processors, or DSPs for short.

Once the signal has been processed by the DSP device it is still in the form of a sequence of numbers and may need to be converted back to an analog signal before being passed to an actuator (for example, a loudspeaker). This process is called digital-to-analog, or D/A, conversion. The complete conversion and processing chain is shown in Figure I.1.

This may seem a lot of trouble for what appears to be a fairly simple task. The misconception here is that signal processing is simple. Although it is possible to do some processing with analog electronics and miss out the conversion stages, the capabilities of analog signal processing are limited. In Chapter 2 we will look in detail at why we process signals digitally, but 50,000 engineers can't be wrong!

# 1

# The Development of Digital Signal Processing

In this brief chapter we hope to explain how mathematical techniques and computer technology have developed to allow us to process real world signals with digital computers. This historical perspective provides some interesting background to today's state-of-the-art in high-performance DSP.

It is not normally appreciated that the development of digital signal processing began because designers of analog signal processing systems wished to simulate their performance before building expensive prototypes. The obvious tool to perform the simulations on was the digital computer and thus began the development of digital signal processing. It is likely that those early pioneers working through the 1950s and early 1960s had little idea that their work would spawn a major area of digital electronics technology for the 1980s and beyond.

Digital signal processing was dependent on the digital computer and the majority of the mathematics or algorithms have thus been developed since 1950. Once these algorithms became established, designers started to look for the computer architectures that would implement them most effectively. The original goal was to produce simulations that would run in an acceptable time. It is not clear when the idea of using digital computers for real signal processing instead of simulations took hold, but once it did, the objectives changed subtly. The ultimate goal became that of real-time DSP, i.e., the system having to complete all required operations in a time short enough for the process being operated on to continue unaffected (see Auslander [1993]). Earlier systems were only able to store the waveform in memory and process it later. Naturally, they could not make decisions based on the data at the time it was changing, hence the term nonreal time.

Our main concern throughout this book is with real-time digital signal processing. The special challenges this presents have stimulated the development of the mathematics and the computer technology applied to DSP at an accelerating rate over the past thirty years.

## ALGORITHMS FOR DSP

The basic mathematical model of continuous signals is based on Laplace and Fourier transforms, which date back to the nineteenth century.

Jean Baptiste Joseph, Baron de Fourier had served as Governor of Lower Egypt under Napoleon, but in 1801 returned to France and continued his scientific researches. In 1822 he published a monumental study on heat flow in which he evolved the Fourier series. The series has since been applied to many branches of science and is one of the principle tools of signal analysis. The Fourier transform (or Fourier integral) is a simple extension of the Fourier series. The basic Fourier series applies to periodic (repetitive) signals and the Fourier transform to aperiodic (nonrepetitive) ones.

Pierre Simon, Marquis de Laplace was a great theoretical astronomer, perhaps the greatest since Newton. Born about twenty years before Fourier, his mathematics was applied to the understanding of planetary motion, but like the Fourier series, the Laplace transform has found wide application in other fields. We are particularly concerned with its modification and application to digital signals. By simple extension and suitable interpretation the Laplace transform becomes a Z-transform. The techniques of the Z-transform are not new and can be traced back to 1730 when De Moivre introduced the concept of the "generating function" in probability theory. The advent of digital computers in the 1940s brought about an increase in the use and application of Z-transforms, which form the basic building block of digital filters. These are covered in detail in Chapter 4.

As with the Laplace transform, the Fourier transform also has an equivalent form for digital signals. The discrete Fourier transform became popular during the 1940s and 1950s as digital computing machines developed, but was computationally demanding. In 1965 an unpromisingly titled paper, "An algorithm for the machine computation of complex Fourier series" by Cooley and Tukey, further refined the DFT. They exploited some special properties and produced a new algorithm known as the fast Fourier transform (FFT), which dramatically reduced the number of multiplications required for calculation. As we shall see, multiplication is a significant bottleneck in the calculation of most DSP algorithms. In reducing the number required by a factor of 100 for large (1024-point) discrete Fourier transforms, the FFT represented a milestone in the development of digital signal processing. There is some interesting literature on the origin of the FFT. Cooley and Tukey brought it into popular use and referred only to the work of Good in 1958 as having influenced them. Other investigations have attributed the origin of the FFT to the German mathematician Runge and even to Gauss, his more eminent earlier compatriot (see Heideman et al. [1984]). Discrete and fast Fourier transforms are covered in detail in Chapter 5.

Recent refinements in DSP theory have on the whole been less dramatic, though research continues unabated into such developments as fuzzy and genetic algorithms, neural networks, etc.

There are several authoritative texts covering the basic theory of DSP with the two "classics" being *Digital Signal Processing* by Oppenheim and Schafer and *Theory and Application of Digital Signal Processing* by Rabiner and Gold. Both were originally published in 1975. These books are detailed theoretical treatments of DSP and should not be tackled by the fainthearted! Throughout our book we shall point you to sources for further reading. We recommend you refer to these advanced texts only when you fully understand the principles we are presenting.

## COMPUTER ARCHITECTURES FOR DSP

When we look at the historical development of the special single-chip processors that are now available for DSP, we should consider device architecture and device technology separately. As is often the case with semiconductor devices, the basic structures were well thought out before the technology was available to support them. Let's first look at the architectural development of DSP devices.

General architectures for computers and single-chip microcomputers fall into two categories. The architecture for the first significant

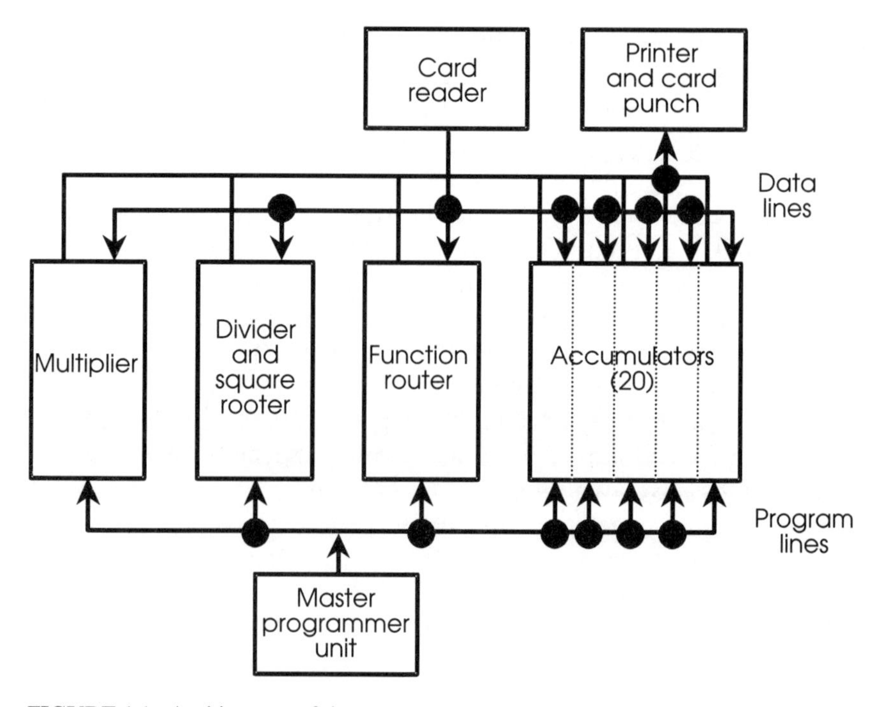


**FIGURE 1.1.** Architecture of the ENIAC – showing separation of data and program memories

electromechanical computer had separate memory spaces for the program and the data, so that each may be accessed simultaneously. This is known as a Harvard architecture, having been developed in the late 1930s by Howard Aiken, a physicist at Harvard University. The Harvard Mark 1 computer became operational in 1944.

The first general-purpose electronic computer was probably the ENIAC (Electronic Numerical Integrator and Calculator) built from 1943 to 1946 at the University of Pennsylvania. The architecture was similar to that of the Harvard Mark 1, with separate program and data memories. This is shown clearly in the block diagram of the ENIAC in Figure 1.1. Due to the complexity of two separate memory systems, Harvard architecture has not proved popular in general-purpose computer and microcomputer design.

One of the consultants to the ENIAC project was John von Neumann, a Hungarian-born mathematician. He is widely recognized as the subsequent creator of a different and very significant architecture, which was published by Burks, Goldstine and von Neumann in 1946 (reprinted in Bell and Newell [1971]). The so-called von

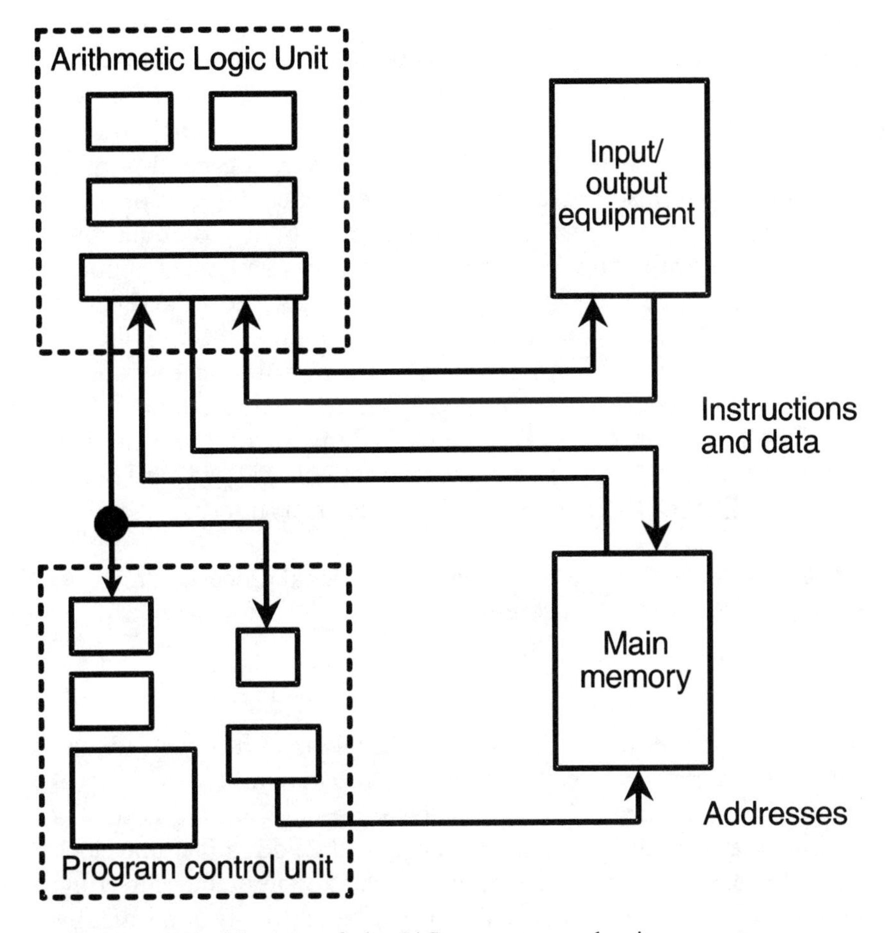


**FIGURE 1.2.** Architecture of the IAS computer – showing common program (instruction) and data memory

Neumann architecture set the standard for developments in computer systems over the next forty years and more. The idea was very simple and based on two main premises: that there is no intrinsic difference between instructions and data and that instructions can be partitioned into two major fields containing the operation command and the address of the operand (data to be operated upon). There was therefore a single memory space for instructions and data.

The structure of a computer built at the Institute for Advanced Studies in Princeton by 1951, and known as the IAS computer, is shown in Figure 1.2. The new architecture simplified the design of the computer, but had the drawback that the machine could only access either the instruction or the data at any one time. History has shown that this limitation is not serious in general-purpose computing.

Common general-purpose microprocessors such as the Motorola 68000 family and the Intel i86 family share what is now known as the von Neumann architecture. These and other general-purpose microprocessors also have other characteristics typical of most computers over the past forty years. The basic computational blocks are an arithmetic logic unit (ALU) and a shifter. Operations such as add, move and subtract are easily performed in a very few clock cycles. Complex instructions such as multiply and divide are built up from a series of simple shift, add or subtract operations. Devices of this type are known as complex instruction set computers (CISC). CISC devices do have a "multiply" instruction, but this will simply execute a series of microcode instructions that are hard coded in on-chip ROM. The microcoded multiply operation will therefore take many clock cycles.

For reasons that will be clear later, digital signal processing involves many calculations of the form:

$$A = BC + D$$

This simple little equation involves a multiply operation and an add operation. Because of its slow multiply instruction, a CISC microcomputer is not very efficient at calculating it. What we need is a machine that can perform the multiply and the add in just one clock cycle. For this we need a different approach to computer architecture. In other words, we need an architecture molded to our application.

In real-time signal processing, our main concern is with the amount of processing we can do before a new item of data arrives that has to be dealt with. Early DSP systems were built using standard components to construct shift-registers, adders and multipliers. The multiplication operation was rapidly seen as the limiting factor in the performance of these computers. Multiplier design advanced through the use of pipelining techniques, and the first single-cycle multipliers were implemented in the early 1970s with standard high-speed emitter-coupled logic (ECL) components.

The leading DSP research institution during this period was Lincoln Laboratories. The Lincoln FDP (Fast Digital Processor) was completed in 1971 and had a multiply time of 600 nanoseconds, but was made up of 10,000 separate integrated circuits! It also suffered from the complications of trying to perform parallel operations using the sequential von Neumann architecture. The Lincoln LSP/2 was built from the lessons learned with the FDP and used an architecture similar in principle to the Harvard 1 computer of the 1940s. By using an inherently parallel architecture, a DSP computer four times faster than the FDP was possible with about one-third as many integrated circuits (ICs).

By the mid-1970s, other leading research institutions had also developed digital signal processing computers with multiply times of around 200ns, all with their own unique features. These machines were certainly capable of real-time signal processing, but were so bulky and expensive that they were not commercially viable. The basic structure of a DSP computer had been created, but it would have to wait for semiconductor technology to catch up. If a DSP computer could be implemented in a few ICs, or even a single device, many commercial avenues would open.

## INTEGRATED CIRCUITS FOR DSP

Throughout the 1970s, integrated circuit technology had become steadily more complex. With the demands of a technology-hungry military machine in the United States and the gradual appearance of electronics in consumer products, there was a great incentive to reduce feature sizes, increase device speed and improve process technologies. The process technology of choice by the end of the 1970s was N-MOS (N-channel Metal Oxide Semiconductor), which worked from a single 5-volt power supply and could be reliably manufactured in a 3-micron geometry, supporting device densities of over 100,000 transistors.

In the first two years of the 1980s, four single-chip digital signal processors became available. The accolade for the first true single-chip DSP is often credited to the American Microsystems Inc. (AMI) S2811, though this is an arguable point. The Intel 2920 and the Nippon Electric Company (NEC) $\mu$PD7720 were also available at around the same time. Slightly later, in 1982, Texas Instruments introduced the TMS32010, and the single-chip DSP had truly arrived.

All early DSP devices use the Harvard architecture to separate data memory and program memory. This allows a program instruction and a data word to be addressed or accessed at the same time. In real-time DSP, the efficient flow of data into and out of the processor is critical. Using a Harvard architecture means that data flow does not need to be interrupted for instructions to be read. The block diagram of the

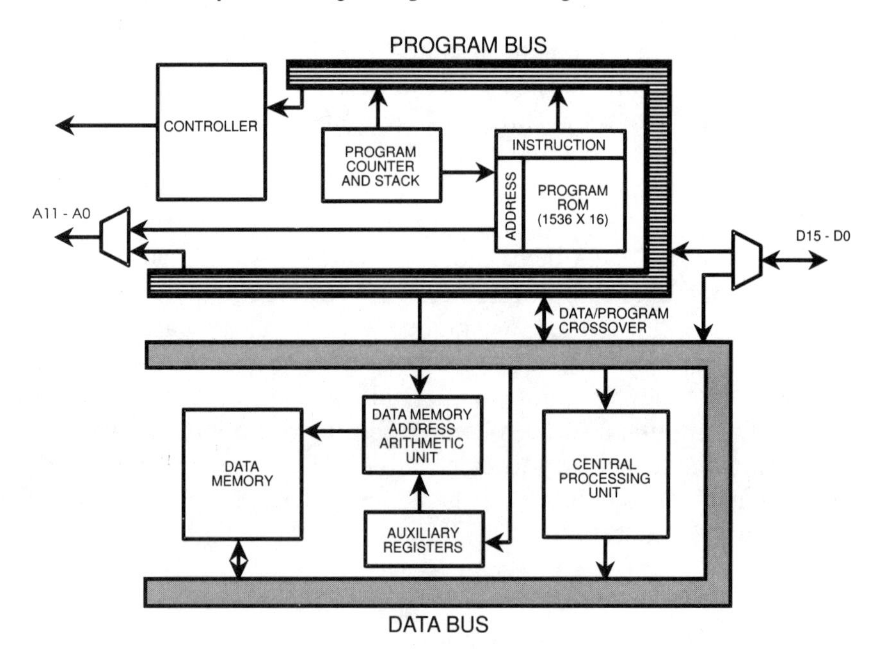


**FIGURE 1.3.** Simplified block diagram of the TMS32010 – showing the separate program and data memory areas of a Harvard architecture

TMS32010 in Figure 1.3 shows the separation of program and data memory and a modification giving a cross-over path between the two memory spaces. This is why the TMS32010 is usually referred to as having a modified Harvard architecture.

By introducing a DSP core onto a device with many standard microcomputer facilities, Texas Instruments moved digital signal processing into a new era. No longer was it almost impossible to design a DSP system and write software for it. The TMS32010 had an assembly language, evaluation tools and an emulator similar to those for any microcomputer device. One level of the "black magic" of DSP was removed. With subsequent developments of faster, more capable devices supported by simulators, debuggers, C-compilers, etc., DSPs have moved into the mainstream of microcomputer system design practice. What is different is the algorithms and the device architectures.

Of course, the story does not end here. The burgeoning computer and personal computer (PC) industry has demanded ever-increasing memory densities. This in turn has led to smaller feature geometries in all integrated circuits and the ability to have more transistors on a single device. Process technologies have improved significantly with the introduction and development of CMOS to the extent that device geometries of 0.35 microns are becoming commonplace. Not only does this allow greater densities of transistors and gates, but it brings with it a reduction in switching delay time for each gate and consequent faster clock cycles and greater processor throughput. In 1994 this means that we see single-chip DSPs of around 4,000,000 transistors with multiply times of less than 40ns for a 32-bit floating-point multiply or as little as 25ns for some 16-bit fixed-point devices.

## REFERENCES

There is a great deal of material that provides information about some aspect of the history of digital signal processing, but we have not found a complete historical review of the subject from the period of De Moivre, Laplace and Fourier onward. Our short discussion is based on Lynn [1982] for the brief biographies of Fourier and Laplace. The Z-transform is described in Jury [1964] and the FFT in various texts. Cooley and Tukey [1965] is the classic FFT paper, but they did not originate the technique. Cooley, Lewis and Welch [1967] is an interesting historical review, attributing the technique to Runge [1903] and the application of his methods to Danielson and Lanczos [1942]. A more recent historical review discusses various advances in numerical analysis in the nineteenth century in England and Germany and attributes an algorithm very similar to the FFT to the German mathematician Gauss (see Heideman et al. [1984]). It's a fascinating subject, but you'll have to look at the literature if you want to find out more.

The development of computer architectures from Babbage through Turing to the modern microprocessor is well documented. The most enjoyable and readable book we have found on the subject is Augarten [1984]. The original von Neumann et al. paper from 1946 may be found reprinted in Bell and Newell [1971]. Other information was from Hayes [1979]. For the later development of DSP computers, useful reviews may be found in Allen [1975] and Bowen and Brown [1982].

Reviews of modern DSP devices are out-of-date almost as soon as they are published, so it is not worth formally referencing any here. Manufacturers' data is a valuable source, as are the occasional review articles in the more authoritative periodicals. The U.S. periodical *EDN* published a major review of then-current DSP devices in its September 29, 1988 edition and has contained other articles since.

We have already mentioned the two classic texts of digital signal processing, but they are worthy of inclusion here as well. Oppenheim and Schafer [1975 and 1988] and Rabiner and Gold [1975] set the standard for general reference works on DSP. They are cited within almost every major publication since. They **are** very theoretical and not suitable for casual reading, but they are the definitive texts on the sub-ject. There have been a number of subsequent texts with a title similar to "Digital Signal Processing", that review the subject in different ways. One of the most useful and interesting that we have found is by DeFatta, Lucas and Hodgkiss [1988], which provides a rigorous DSP system design methodology for all the steps leading up to hardware and software production.

There is a vast and increasingly bewildering array of recent books on DSP. Many of these specialize in some particular aspect and some of the best are referenced in subsequent chapters.

Allen, J. [1975]. "Computer Architecture for Signal Processing," *Proceedings of the IEEE*, vol. 63, no. 4, pp. 624–633, April 1975.

Augarten, S. [1984]. *Bit by Bit,* Ticknor & Fields, New York.

Auslander, E. [1993]. "Digital signal processing and the emerging markets of the '90s," *Le Traitement du Signal et Ses Applications, Actes des Conferences,* DSP'93.

Bell, C.G. and Newell, A. [1971]. *Computer Structures,* McGraw-Hill, New York.

Bowen, B.A. and Brown, W.R. [1982]. *VLSI Systems Design for Digital Signal Processing, Volume 1: Signal Processing and Signal Processors,* Prentice-Hall, Englewood Cliffs, NJ.

Cooley, J.W.; Lewis, P.A.W. and Welch, P.D. [1967]. "Historical Notes on the Fast Fourier Transform," *IEEE Transactions on Audio and Electroacoustics,* Vol AU-15, No. 2, pp. 76–79, June 1967.

Cooley, J.W. and Tukey, J.W. [1965]. "An algorithm for the machine computation of complex Fourier series," *Math. of Comput.*, Vol 19, pp. 297–301.

Danielson, C.G. and Lanczos, C. [1942]. "Some improvements in practical Fourier analysis and their application to X-ray scattering from liquids," *J. Franklin Inst.,* Vol 233, pp. 365–380 and 435–452, April 1942.

DeFatta, David J.; Lucas, Joseph G. and Hodgkiss, William S. [1988]. *Digital Signal Processing: A System Design Approach,* John Wiley, New York.

Hayes, John P. [1979]. *Computer Architecture and Organization,* McGraw-Hill International, New York.

Heidemann, Michael T.; Johnson, Don H. and Burrus, C. Sidney [1984]. "Gauss and the History of the Fast Fourier Transform," *IEEE ASSP Magazine*, pp. 14–21, October 1984.

Jury, E.I. [1964]. *Theory and Application of the Z-Transform Method,* John Wiley, New York.

Lynn, Paul A. [1982]. *The Analysis and Processing of Signals,* MacMillan, London.

Oppenheim, A.V. and Schafer, R.W. [1975 and 1988]. *Digital Signal Processing,* Prentice-Hall, Englewood Cliffs, NJ.

Rabiner, L.R. and Gold, B. [1975]. *Theory and Application of Digital Signal Processing,* Prentice-Hall, Englewood Cliffs, NJ.

Runge, C. [1903]. *Zeit. fur Math. and Physik,* Vol 48, p. 433.

# 2

# Why Do It Digitally Anyway?

By now you have read about what DSP is and how it has developed over the last thirty years, but still know nothing about why you should use it. After all, it would be possible to use analog circuits to perform most of the work done by digital signal processors. Despite this, there has been a vast amount of money invested by semiconductor companies in developing ever faster chips for digital signal processing and an ever more eager uptake of these devices by designers. So just what is it about DSP that makes it so popular?

The answer is not short and clear-cut. There are many advantages in using digital techniques for general-purpose signal processing. These advantages also apply to more specific applications, but there are also some functions that may be performed by DSP that cannot be implemented in analog systems. The advantages of using digital techniques in signal processing fall into several broad categories: programmability, stability, repeatability, easier implementation of adaptive algorithms and the ability to implement error correcting codes and special functions such as linear phase filters. We may also consider data transmission, storage and compression as areas where digital processing has significant advantages over analog.

Looking at this list, it seems that DSP is the universal panacea. Unfortunately, life isn't that simple. There are limits to what can actually be done by realizable DSP systems, and there are even some areas in which an analog solution is preferable. In this chapter we will look at the advantages and capabilities of DSP. We will also look specifically at the area of control systems, where DSP can offer an improvement not only over analog control, but also over traditional microcontroller-based solutions.

## PROGRAMMABILITY

One reason for the universal uptake of the digital computer (in the form of PCs) is that they are programmable and reprogrammable. We take this for granted and can use the feature to turn our PCs from word processors to games consoles in a few seconds. Microprocessor technology gives the same advantage to a digital signal processing system.

It is possible to design one hardware configuration that can be programmed to perform a very wide variety of signal processing tasks, simply by loading in different software. For example, a digital filter may be reprogrammed from a low pass to a high pass with no change in hardware. In an analog system the whole design would need to be changed.

In many cases there is no need to reprogram a system, only to upgrade its operation. Examples might include missile guidance systems, where use in combat circumstances may highlight deficiences not found in trials. The ability to perform these alterations by the simple replacement of a single memory device is a significant advantage for DSP. To the extent that one can upgrade an analog signal processing system, it is done by changing component values – generally a soldering iron job. There is a limit to the amount that its function or performance can be changed like this.

## STABILITY

When we look at the field performance (i.e., how systems perform in use over a period of time), the situation gets even worse. Components including resistors, capacitors and operational amplifiers all change their characteristics with changes in temperature. This means that an analog circuit may perform quite differently at 0°C than it does at 70°C. Again, digital circuits will show no variation with temperature throughout their guaranteed operating range.

A third form of variability that affects analog circuits is component aging. Capacitors in particular are prone to aging of the dielectric material. This will cause a change in impedance and alter the behavior of the circuit. Compensations have to be built into the circuit to allow for component, thermal and aging variations. This can greatly complicate the design process and compromise overall circuit performance.

Putting all this in perspective, there are many thousands of analog systems in use for signal processing, so the problems are by no means unsurmountable. Nevertheless, they do exist and the use of a digital signal processor removes these variables. Furthermore, DSP circuits may even be programmed to detect and compensate for changes in the analog and mechanical parts of a complete system.

## REPEATABILITY

Digital systems are inherently repeatable. If you ask five hundred identical digital computers to perform a sequence of sums, they will all provide the same answer, exactly the same answer. If you apply a signal to five hundred analog circuits, built using components of identical specification, you will certainly not get the same output from each circuit.

The reason for this is very simple – there is a spread of performance of components in analog systems. Resistors have a tolerance specification, usually 5% of their value, but more expensive components may have a 2% or 1% tolerance. Typical capacitors have a tolerance of 20% or worse. Similarly, analog (or linear) semiconductor circuits will have a specification range within which they are guaranteed to operate, but natural variations in the manufacturing process will mean that their precise performance varies from device to device. This means that it is impossible to predict the precise behavior of an analog system. The consequence of this is that, if precision operation is required, it is necessary to adjust the performance of each system built by including variable resistors or capacitors into the system design and adjusting or "tweaking" them during a calibration test.

## EASIER IMPLEMENTATION OF ADAPTIVE ALGORITHMS

Several years ago, DSP systems were developed that could cancel some of the noise within the cockpit of a car, helicopter or air-plane.

In the case of the car, the noise that is cancelled is originally caused by the engine and the resonances set up in the body panels by engine vibrations. The noise cancellation system takes the engine speed as a reference and attempts to produce an "anti-noise" signal to cancel the cockpit noise. There are microphones in each headrest that determine the success of the attempt. Based on the changes detected by the microphones, the system changes the characteristics of the anti-noise until the best noise reduction possible is achieved. When the engine speed changes, the system adapts once more to the new engine speed.

A DSP system can easily adapt to some change in environmental variables. The adaptive algorithm simply calculates the new parameters required and stores them in memory, overwriting the previous values. Some very basic level of adaption is possible in analog systems, but the complete change of a complex set of filter characteristics (as used in noise cancellation) is beyond the practical scope of analog signal processing.

## ERROR CORRECTING CODES

Applying error detection and correction when retrieving or transmitting data has become vital. Compact discs and data modems demonstrate the scope of the application of error detection and correction codes.

Data retrieval and data transmission systems suffer from a number of potential forms of error. In the case of the compact disc, it is surface damage, manufacturing defects, or mechanical misalignment. In the case of a data modem, it may be noise or echo on the line, especially where that is an analog telephone line.

With information in a digital or binary form, we may easily build into the data stream additional "redundant" bits that are used to detect when an error has occurred in the principal data. In more sophisticated systems, algorithms are applied to the generation of the redundant bits that will allow the original principal data to be reconstructed. The first case may be nothing more than parity checking; the second will involve more complex techniques such as block codes and error correction. It's a highly technical subject that is covered to some extent in most texts on digital data transmission (see Chapter 6). Accessible specialist texts are not so easy to find, but Arazi [1988] is a thorough introduction.

## SPECIAL FUNCTIONS

There are some valuable signal processing techniques that cannot be performed by analog systems. The classic example is that of linear phase filters. A digital finite impulse response filter with conjugate even symmetry of coefficients about the mid-point will possess a linear phase response with frequency. Another type of filter that may easily be implemented digitally is a notch filter with a steep cutoff. It is virtually impossible to create similar analog filters. There is a full explanation of digital filters in Chapter 4, including a detailed study of different types of filter and their uses.

Control systems provide further examples that can only be realized with digital techniques. One example is known as a deadbeat controller. These are used in situations where a very rapid settling time is required. Digital controllers are also able to infer velocity from the output of an encoder, so a separate tachometer is not necessary. This obviously reduces system cost and can increase reliability.

We have already mentioned lossless data compression in digital data transmission and storage. Lossless compression is not possible with analog signal processing.

## DATA TRANSMISSION AND STORAGE

Although some "experts" may not agree, for the majority of people, the onset of the compact disc (CD) player brought trouble-free high-quality audio into the home. The black vinyl disc with its clicks, pops, and hiss is gradually being consigned to the museum. It is clear that the fidelity of the digital medium is greater than that of the analog one. The ability of record-playing systems to introduce gross errors into the music (e.g., clicks and pops) is well known. Unless a CD player receives a severe mechanical shock, there will be relatively few errors, and those that there are can be corrected digitally, with no loss of fidelity.

We are considering here a number of factors such as: signal-to-noise ratio, error protection, detection and recovery, data compression, etc. The subject is worthy of a chapter or even a whole book of its own. We will simply give a brief overview of some easily understood and readily observed differences between analog and digital signals.

Continuing with the example of domestic hi-fi systems, let's look at signal-to-noise ratio. The term is self-explanatory. What is important is the maximum amount of signal for the minimum amount of noise. If we look at the music reproduction process from disc to power amplifier, we can compare the noise performance of digital and analog audio.

Taking the digital system first, there will be some level of noise on any compact disc, caused by something called quantization error. This is explained in detail in Chapter 3 and we needn't worry about it now. Once the disc is inside the player, a laser is used to read the digital stream of information from the disc surface. There will be noise introduced at this stage in the form of data errors caused by poor disc manufacture, disc damage, mechanical shock to the laser, or poor alignment. Because of a special code added to the music information, many of these errors can be corrected by a special DSP device inside the player.

In simple terms, the music signal on the original CD has a 16-bit word length. In order to maintain accuracy, most intermediate calculations are stored to at least 32-bit accuracy. With appropriate oversampling, a 20-bit value may be presented for conversion to the analog waveform. There is effectively no loss of information caused by the digital processing of the signal.

We can also discount noise caused by a corruption of the digital information. In digital circuits, there are only two possible states, a logic "1" or a logic "0". In 5-volt systems the difference between the two is typically a minimum of 3 volts. The only way in which the signal can become corrupted while being transmitted in its digital form is by the introduction of drastic noise, causing what should be a logic "1" to be interpreted as a logic "0" or vice versa. This does not happen in any competently designed digital circuit.

Once all the processing is complete, the digital music signal is passed to a special circuit that converts it to an analog signal, so that it may be amplified conventionally. The digital-to-analog conversion process does not re-create a perfect analog waveform, and some analog filtering is usually required before the signal can be passed to an amplifier. This analog filtering stage is regarded as undesirable, and many techniques such as interpolation or oversampling are implemented by DSPs within the CD player to reduce or eliminate the need for analog filtering. The output from the digital-to-analog converter is typically up to 2 volts in amplitude and requires only a passive volume control and power amplification. Digital-to-analog conversion is covered in detail in Chapter 3.

Now we turn to the ubiquitous vinyl long-playing record (LP). The manufacturing process involves the use of submasters for pressing vinyl discs. These wear during use and have a finite life. Clearly, small differences from the original metal master can be introduced at this stage. There are further problems with the vinyl material used for LPs which is very prone to damage by handling, a faulty pickup stylus or poor storage conditions. There is also a tendency for the surface to attract electrostatic charge, which may discharge at any time through the stylus. There is no error detection or recovery mechanism for these faults and the subsequent noise generated.

There is a vast array of different cartridge designs available for converting the mechanical signal (grooves) on the record into electrical energy. Even a relatively untrained ear can detect significant differences between cartridge models. The accuracy with which the conversion is made is clearly questionable. Typical cartridges produce an output of 1–2 millivolts. In the case of the cheaper moving-magnet cartridges this output is nonlinear with frequency and requires both equalization and amplification.

The signal from the cartridge is carried by a coaxial cable to the equalizer/amplifier. Coaxial cable provides good (though not perfect) isolation from external interference. Further noise will inevitably be introduced within the equalizer/amplifier through the power supplies to the circuit, signal crosstalk, inaccurate equalization, and so on.

The clear message that comes over from this simplified comparison is that digital information is more robust than analog information. There are some issues concerned with the conversion process from analog to digital and back again, but by careful attention to simple principles, these can be minimized. See Chapter 3 for a detailed discussion of the conversion processes and some practical examples of different approaches to the subject.

## DATA COMPRESSION

The reasons why compression of speech, images and other data is important are that information channels cost money and transmission bottlenecks often make compression necessary for real-time processing constraints. Satellites, optical fibers and cables are all expensive to install and maintain. The aim therefore is to get the maximum of information transferred in the minimum amount of time.

In all examples of analog compression some information is lost. A typical example is the bandwidth limiting applied to analog telephone lines in order to multiplex calls. This effectively limits the frequency response to 3kHz.

With digital data transmission or storage, there are two forms of compression: lossless and lossy. In lossless compression, when the information is restored it is unchanged from the original prior to compression. In lossy compression there is some level of loss of information, normally in the fine detail. In speech and image compression, some degradation in quality may be quite acceptable to the viewer or listener. When financial or other data is involved, obviously no change can be allowed. Lossy compression has the advantage that a greater level of compression is possible.

Data modulation and compression techniques are covered in more detail in Chapter 6.

## PRACTICAL DSP SYSTEMS

We must temper some of this enthusiasm for digital signal processing by pointing out that there are limits to what can practically be done in DSP. Although new, faster DSP devices are regularly introduced, there is still a limit to the processing that can be done in real time. This limit becomes even more apparent when system cost is taken into consideration. New developments in parallel processing for DSP have moved its capabilities into new performance areas, but only where the increased cost can be justified.

So what can actually be done with DSP and at what cost? We have already looked at some examples of DSP applications: the compact disc (CD) player, adaptive noise control in cars, etc. Discount-store prices of well under 100 dollars for a CD player demonstrate that DSP is not an expensive technology. As a further illustration, the DSP device required for the adaptive noise control system would cost under 10 dollars in volume and the lowest-cost single-chip DSPs available cost around 3 dollars in volume. Even these devices are capable of performing simple speech recognition or the signal processing required in a V.22bis (2400 bits per second) modem.

For an example of what may be done with existing DSP devices, we will look at the five generations of the Texas Instruments (TI) TMS320 family. It is not necessary to dwell on the features of each generation at this point. For now, just remember that the first (TMS320C1x), second (TMS320C2x) and fifth (TMS320C5x) generations are 16-bit fixed-point devices, and the third (TMS320C3x)

**TABLE 2.1 Some typical applications for TMS320 family DSPs**

| TMS320C1x | TMS320C2x | TMS320C5x | TMS320C3x | TMS320C4x |
|---|---|---|---|---|
| DTMF decoders | Active suspension | GSM vocoder and equalizer | MPEG audio codec | RADAR |
| ADPCM and LPC codecs | Digital servo motor controllers | IS54 and JDC vocoders | LD-CELP vocoders | Virtual reality animation |
| Musical instruments | Vocoders | ADPCM with echo cancellation | Videophones | Parallel processing systems |
| Toys | Tapeless answering machines | V.3x modems . . . . . | Voice mail systems | Image recognition . . . . . |
| Hard disk drive controllers | FAX machines . . . . . | | Hi-Fi systems | |
| V.2x modems | | | 3-D graphics accelerators . . . | |

and fourth (TMS320C4x) generations are 32-bit floating-point devices. Price points are not possible to indicate accurately in a textbook, but it is important to recognize that these are highly cost-effective devices, and typical TI fixed-point devices cost much less than one dollar per MIPS (million instructions per second). Table 2.1 shows some typical commercial applications for each generation of the TMS320 family.

Although this list is impressive, there are limitations to what single-chip DSPs can do. We can't yet have infallible voice-controlled access or speaker independent voice recognition, though for how much longer? The TMS320C40 has brought true parallel processing to the DSP world, so almost anything is possible. The drawback is that there is still a close link between cost and system performance, with many applications simply not being economical at present. However, as semiconductor technology continues to become more efficient in terms of performance per dollar, the applications it makes sense to build increase in number and scope all the time.

The penetration of DSP into consumer goods has helped bring down the price of DSP devices. Equally the development of parallel processing DSPs has allowed far higher performance DSP systems than ever before. The scope of DSP applications is thus broadening at both ends of the performance scale, one being driven by low component cost and the other by high component and system functionality.

## DSP IN CONTROL SYSTEMS

An increasingly important area of application for DSP is in control systems. There are many features of digital systems that make them desirable for control. Digital signal processors have now brought the kind of sampling rates and real-time performance that can widen the scope of digital control.

Despite the growing popularity of digital control, analog solutions have persisted longer than in other areas of signal processing. The simple reason for this is that control applications are different from general-purpose signal processing. Analog systems give effectively infinite sampling rates and are infinitely variable. Digital systems must have a finite sampling rate and a finite number of output levels. This has important implications for control system performance and stability. However, the limitations of digital control may be overcome with careful design, and many TMS320 family devices are now used in a wide variety of control systems.

The theory of digital control is rather different from that of general-purpose digital signal processing and we do not intend to go into it in any depth in this book. There are a few specialist texts available and a number of excellent papers that provide an introduction to the subject. These are described at the end of the chapter. It is also planned to produce a further volume in this series that deals specifically with the implementation of control systems using digital signal processors.

## THE INCREASING CAPABILITIES OF ANALOG VLSI

Having spent a great deal of time looking at the advantages and capabilities of digital signal processing and processors, it is only fair to provide a balance.

The most significant performance improvements in electronic systems can certainly be attributed to the availability of extensive digital processing power made economically viable by process enhancements and geometry reductions. Nevertheless, there have been major advances in analog processing techniques that still require the system architecture and partitioning to be carefully considered.

The euphoria and extrapolation of the emergence of digital technology in the early 1980s made it look like the only analog components that would soon be needed would be an "ideal" operational amplifier (op amp) and an accompanying multibit analog-to-digital converter (ADC) with no conversion errors. With its infinite gain bandwidth, zero noise and impeccable dc characteristics, the op amp would be configurable to amplify and equalize the signal originating from any transducer. With its zero conversion time and umpteen-bit resolution, the ADC would present the signal uncorrupted to the digital signal processor where it could surely come to no further harm.

Silicon real estate costs being what they are dictates that the end function costs be evaluated in all possible implementations—analog, digital or mixed signal (a combination of analog and digital on a single die). In some applications advances made in analog components will make them the favored approach. They do not display any of the limitations of digital systems such as quantization effects and so on (see Chapter 3).

## PROGRESS TOWARD THE IDEAL OP AMP

The parameters that have seen the most substantial improvements over the last decade have been:

dc accuracy (offset voltage, bias current, etc.)
Speed (slew rate, bandwidth)
Noise
Power consumption
Price/performance

To a great extent the ideal op amp is here, though it resides in more than one package! By this we mean that all of the shortcomings have been addressed to an extent that they hardly fall short of idealized components but are features of more than one device type. This is rarely an issue since users are often concerned with only a few parameters. The circuits are therefore configured to address application areas that are primarily interested in one set of specifications. These will then be addressed by particular technologies and circuit techniques.

Input offset nulling techniques have allowed the introduction of devices such as TI's TLC2652 with an input offset voltage of $1\mu$V. Guaranteeing a $1\mu$V offset voltage maximum requires scrupulous control of the tester since thermal errors might easily swamp the actual amplifier-induced error.

Bias current is composed of several components. The package leakage itself can be the most substantial and much care is taken in modern assembly plants to minimize this. Second, the structures used to protect the transistors from electrostatic discharge (ESD) must be carefully chosen to avoid worsening the input current. Modern CMOS analog processes have a virtually unmeasurable gate leakage current and special passivation layers ensure this performance does not deteriorate during the life of the device.

Modern processes for analog components have benefited from the advances made in the achievable geometries of transistors, as have digital components. The smaller a transistor can be made, the smaller the parasitic capacitances and resistances can be made. This enhances the speed of the transistors. The circuits they are used in can therefore be designed for more broadband use. This has particular benefits in telecommunication circuits and in TV and video products.

Most of the advances in reduction of noise in analog components have been due to process control. Random effects of "popcorn" noise affecting apparently identical batches of circuits are a thing of the past. These effects were mainly traceable to sodium contamination inducing surface states. Performance limitations are now practically the immovable ones associated with fundamental laws of physics. The characterization of processes for noise is now much more thorough and sophisticated. Utilizing transistors of optimum geometries to minimize their noise contribution is a well-known science, and not the black art it was in the past.

Virtually every new application for electronic components has a premium attached to the power consumption of the circuitry. Consider a mains metering circuit consuming 1mA. This per se is not an issue, however the power is consumed "down-side" i.e., paid for by the supply company, not the consumer. In the UK there are currently 14 million meters. If each consumes 1mA at the rated 240V we have a total loading of 3.36MW. This is the capacity of a large power station just for the meters.

Techniques using sampled sensing are becoming very popular. The idea is to wake the detection circuitry briefly from its quiescent state to do an accurate measurement with a very low duty cycle. This has the advantage that a "quality" reading may be taken using enough current to get low noise performance while keeping overall averaged current consumption low. Circuits have been devised to come awake and take a 12-bit reading and turn off again in under 5$\mu$s.

Overall, there has been a substantial reduction in the price/performance trade-off for all analog functions. This is particularly true in professional-grade standard components like op amps, comparators, data converters, etc. The availability of truly premium specification components at an affordable price has provided performance and accuracy not previously commercially viable. Scale of manufacture, low-cost packaging including surface mount and test cost reductions have all made a contribution.

## OTHER CIRCUIT FUNCTIONS

**Analog-to-digital converters:** The different techniques for producing ever faster and higher resolution are covered in the next chapter. The sigma delta converter is the focus of much attention at the moment and uses digital techniques extensively as we will see. There is a danger that some other methods such as voltage-to-frequency converters will be overlooked. They provide the accuracy of a sigma delta device, but much more slowly. They have the often-overlooked advantage that their dc performance is far greater.

**Comparators:** Almost all of the techniques and process enhancements applied to op amps are also applicable to comparators. They have benefited from the auto-zero, CMOS, current mode, low-power and low cost approaches exploited to improve op amps and are similarly close to the ideal component for many applications. Their speed is a distinguishing feature however. Here it must be pointed out that not much in the way of fundamental advance has been made in the last decade. Speed and power consumption are still inextricably linked, and there has not been a major improvement in their figure of merit even in the last few years.

## SUMMARY

Significant improvements have been made in analog design over the past ten years. This period has also seen the massive increase in the use of DSP devices and systems to solve a whole series of problems. The advantages of digital signal processing are very powerful and it is unlikely that there will be any great resurgence of general-purpose analog signal processing in the near future. We are certainly seeing analog circuitry being used extensively in partnership with DSP in the same design, often to provide some basic preprocessing function.

It is also clear that DSP has bandwidth limitations related to processor cycle times and algorithm complexity. Analog circuitry can be applied to far higher bandwidth signals. This will continue to drive a demand for ever better analog devices in applications such as mobile communications. In both the analog and digital domains it is certainly true that any new technology advances are eagerly adopted by users and new system demands continue to stretch device designers. We can look to mobile communications, high-definition TV, videophones, etc., to provide significant challenges for analog and digital components and for data converters for years to come.

## REFERENCES

There are many references that put forward the general advantages of digital signal processing compared to analog. Most of these are papers from manufacturers of DSP devices and do not go into any greater depth than we have done.

Although we do not cover the subject in this book, there is a growing interest in the use of digital signal processors in control systems. Indeed, it is hoped that an introduction to the subject will be published as one of a series of books to follow on from this one. There are a great number of books concerned with the general subject of digital control and no shortage of theoretical studies of Kalman filters, etc. What is more difficult to find are books and papers that are directly concerned with the advantages of using DSP devices in digital control systems. We have listed a few references below. The textbooks by Dote [1990] and Lewis [1992] are fairly advanced but do include examples for TI's TMS320C25 fixed-point DSP device. The book by Ahmed [1991] is a collection of papers, some specially commissioned.

Ahmed, Irfan (ed.) [1991]. *Digital Control Applications with the TMS320 Family,* Texas Instruments, Dallas, TX.

Arazi, Benjamin. [1988]. *A Commonsense Approach to the Theory of Error Correcting Codes*, MIT Press, Cambridge, MA.

Dote, Y. [1990]. *Servo Motor and Motion Control using Digital Signal Processors,* Prentice-Hall, Englewood Cliffs, NJ.

Hanselmann, H. [1987]. "Implementation of Digital Controllers - A Survey," *Automatica,* Vol 23, No 1, 1987.

Lewis, F. [1992]. *Applied Optimal Control & Estimation: Digital Design & Implementation,* Prentice-Hall, Englewood Cliffs, NJ.

# 3

# Converting Analog to Digital

We have seen that signals in the real world are analog. In this chapter we shall first look at how we convert them into digital patterns that can be used by a DSP and then how we rebuild the analog output signal.

The first part of the process that we shall look at is called *sampling*. To illustrate this operation we shall use the changing weight of a baby, as shown in Figure 3.1. The information was actually gathered from the fortnightly visit to the clinic where the baby was weighed, as shown in Table 3.1. Plotting these values against time (Figure 3.2), we see we have a set of snapshots of the baby's progress. These snapshots are called *samples* of the input signal (the baby's weight). In other words, sampling an input signal is a method of recording an instantaneous value of that signal.

There are many other examples that we can use. We often see daily figures for rainfall, the number of hours of sunshine and the depth of snow in ski resorts. What is notable is that the sampling period is always a fixed unit: fortnightly for the child and daily for the rain, sunshine, and depth of snow. This allows us to understand the information given and to possibly make a decision on what to do next.

Let us take another example: A stockbroker wants to predict what is likely to happen to his portfolio of shares. We shall assume that the stock index is going to vary continuously over time, as shown in Figure 3.3. Also assume that the stock index was published only at the

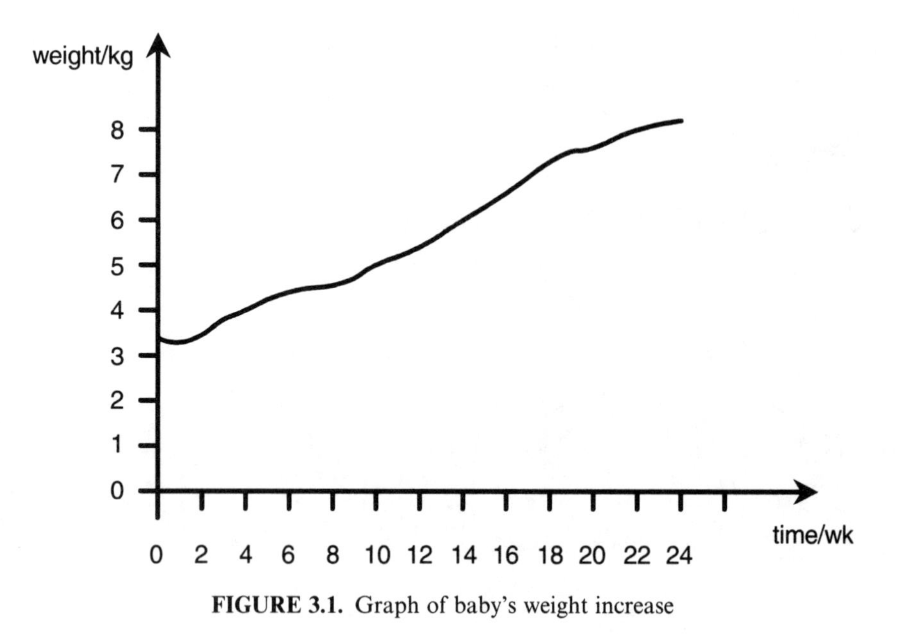


**FIGURE 3.1.** Graph of baby's weight increase

irregular times $t_1$ to $t_7$ shown in Figure 3.3. We can plot these published values against time (Figure 3.4), but as we have no information other than the stock index we must infer what happened between the published values. The easiest way to do this is to just join up the values with a simple curve. The implication of this is that we

**TABLE 3.1 Baby's weight chart**

| Week | Weight (kg) |
|---|---|
| 0 | 3.45 |
| 2 | 3.50 |
| 4 | 3.95 |
| 6 | 4.49 |
| 8 | 4.59 |
| 10 | 4.98 |
| 12 | 5.37 |
| 14 | 6.05 |
| 16 | 6.70 |
| 18 | 7.41 |
| 20 | 7.52 |

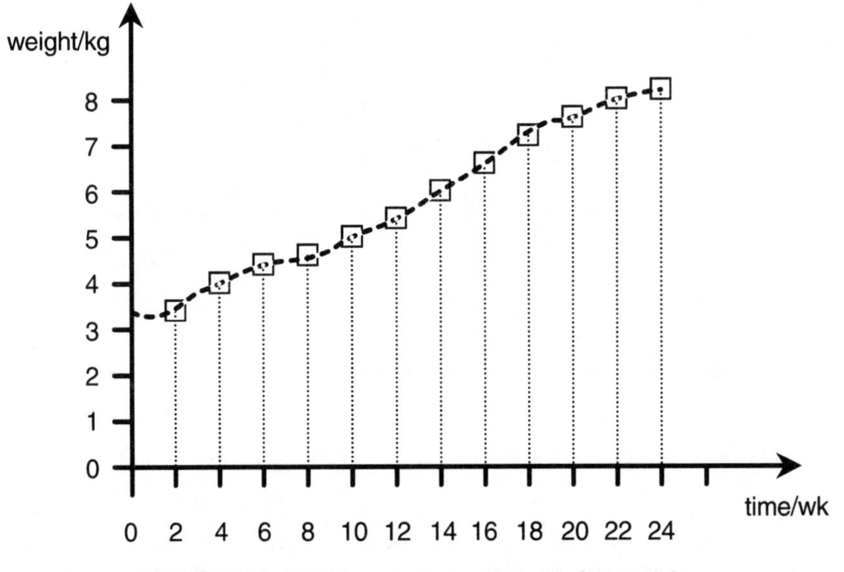


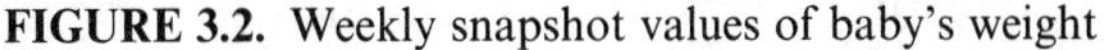

**FIGURE 3.2.** Weekly snapshot values of baby's weight

don't see the dip in the stock index that actually occurred between $t_4$ and $t_5$. We have missed what may have been an important signal to the collapse of share prices. Furthermore, we missed the brief drop in share values below our selling price threshold ($P_{sell}$).

The critical element in accurately capturing the meaningful information in the analog event is the frequency at which the snapshots are

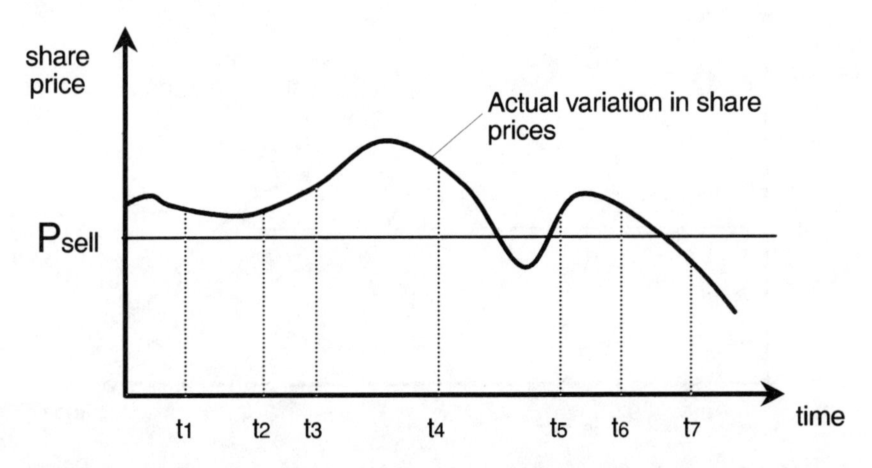


**FIGURE 3.3.** Actual variation of share price with time and publication times ($t_1$ to $t_7$)

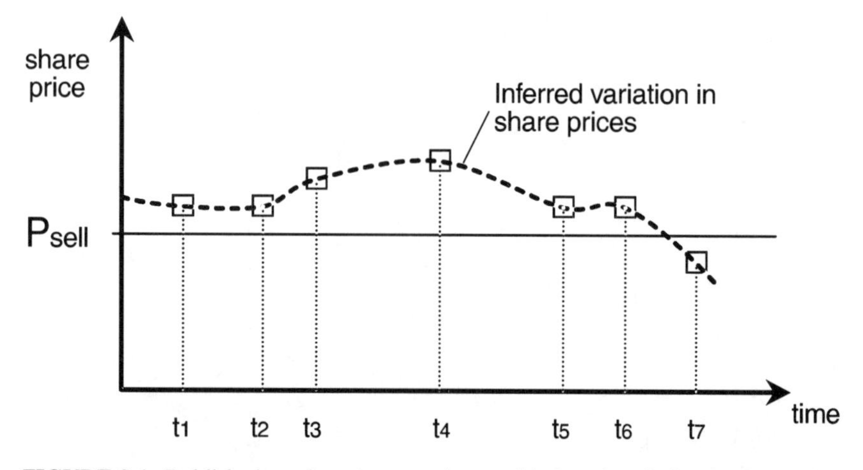


**FIGURE 3.4.** Published stock exchange values and inferred variation in share prices

taken. If we look again at the example of the stockbroker, even if we sampled at regular intervals we could still miss the crucial dip if the sample period is too long (Figures 3.5 and 3.6). The only way to guarantee that we do not miss the dip in value is to sample at a greater frequency.

The original theorem that defines the minimum frequency required to accurately represent an analog signal, or more precisely a continuous band-limited signal (as in Figure 3.7b), is called the Nyquist Sampling Theorem. Jerri [1977] reviews various extensions to

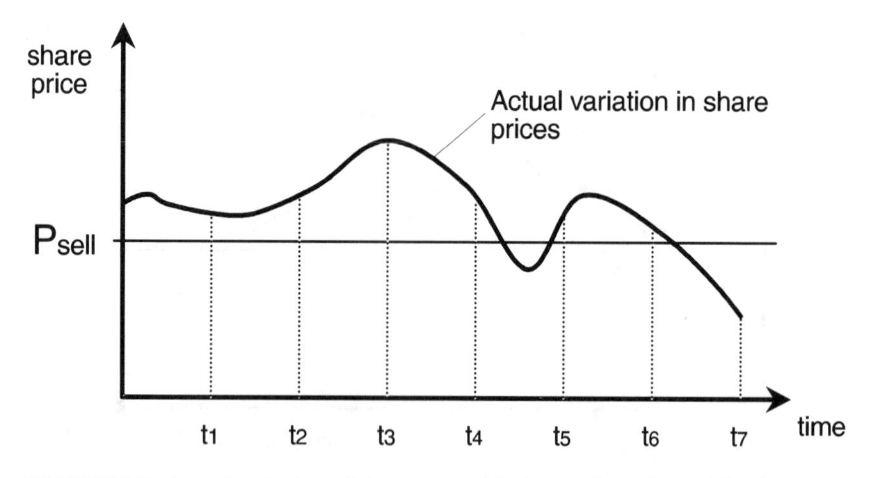


**FIGURE 3.5.** Actual variation of share price with time and regular publication times ($t_1$ to $t_7$)

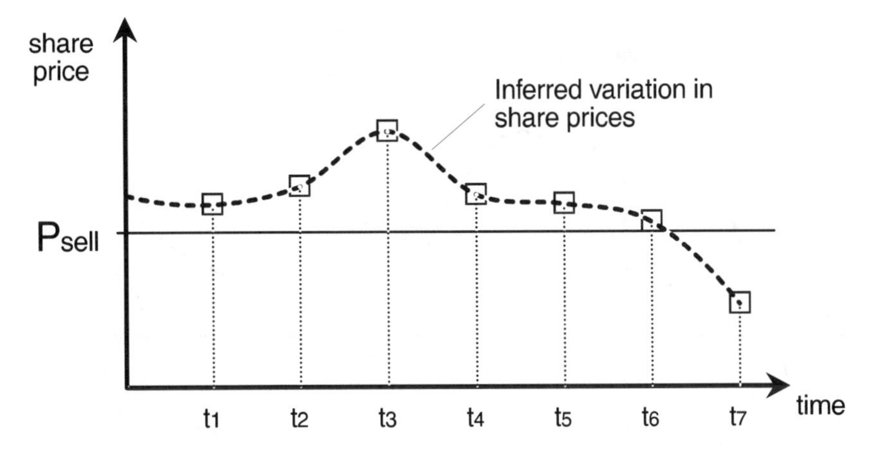


**FIGURE 3.6.** Regular published stock exchange prices and inferred variation in share prices

Nyquist's work, some of which are actually based on even earlier work. The most commonly stated theorem in modern texts is Shannon's Sampling Theorem.

## SHANNON'S SAMPLING THEOREM

If a function $f(t)$ contains no frequencies higher than W cps (Hz) it is completely determined by giving its ordinates at a series of points spaced (1/2W) seconds apart. In other words, in order to accurately represent an analog signal, the minimum sampling frequency must be equal to or greater than twice the highest frequency component of the original signal. This minimum sampling frequency is often referred to as the Nyquist frequency or Nyquist limit.

To be strictly accurate we should use the term *bandwidth* rather than highest frequency component. The reason for this is not intuitively obvious, but relies on mathematical explanation. Jerri [1977] provides adequate references for further reading on this subject. We shall now provide an intuitive description of Shannon's original sampling theorem, which will be more than adequate for most purposes.

The effect of sampling the input signal (Figures 3.7a and 3.7b) is similar to multiplying the signal with a series of sine waves at multiples of the sampling frequency, $f_s$. This means sampling is actually very similar to analog modulation – the input spectrum is repeated every $f_s$ (Figure 3.7c). We shall use this property to deduce Shannon's theorem.

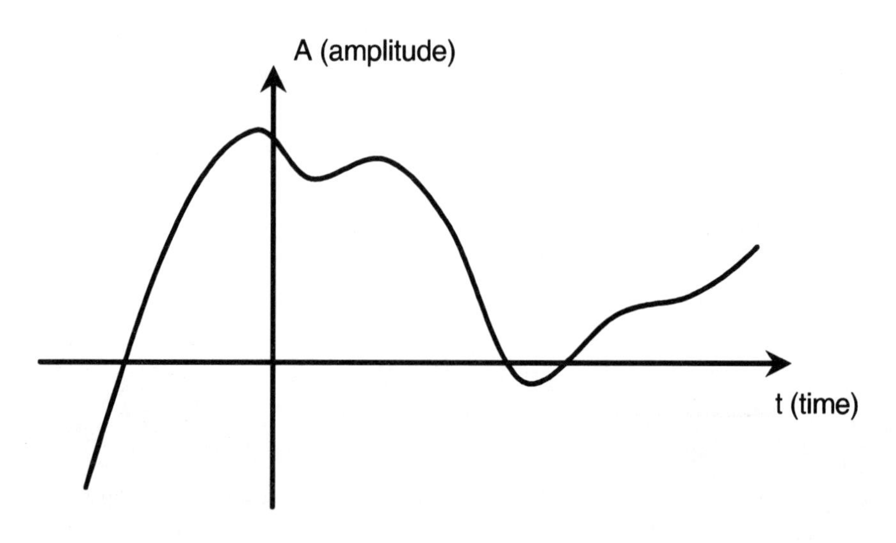


**FIGURE 3.7a.** Analog signal to be sampled

Consider the effect if $f_s < 2f_m$, where $f_m$ is the maximum frequency of the input signal (Figure 3.7d). The frequency bands overlap, which causes interference in the output signal. This effect is called *aliasing*. Don't worry too much if this isn't crystal clear. The next section should at least show you the reasons why aliasing is a problem.

The effect of aliasing on the input signal can be demonstrated using different sampling rates on a sine wave of frequency $f_a$, as is shown in Figure 3.8a. In Figure 3.8b we are sampling at the Nyquist

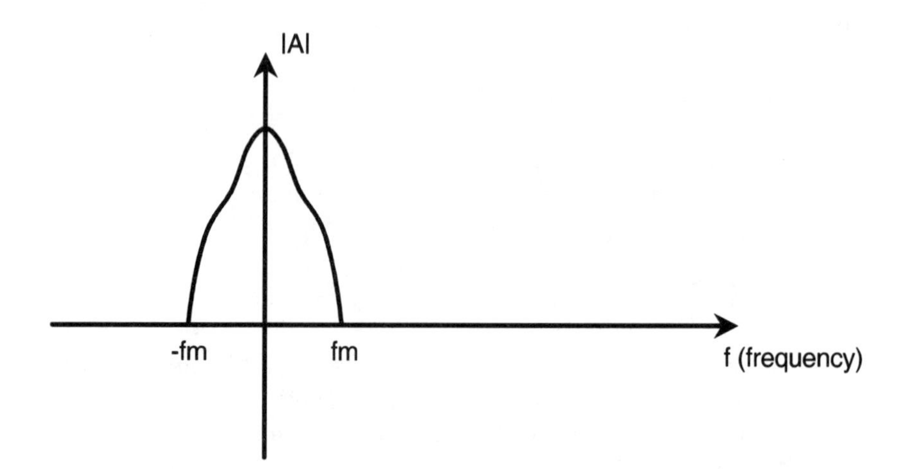


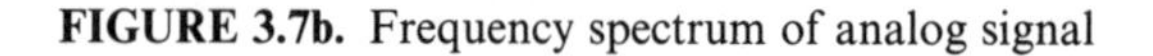
**FIGURE 3.7b.** Frequency spectrum of analog signal

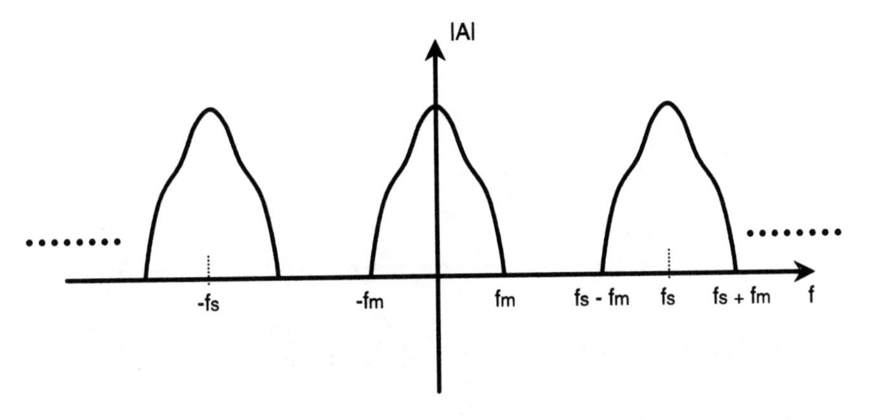


**FIGURE 3.7c.** Sampled spectrum: $f_s > 2f_m$ – spectrum repeats every $f_s$

limit ($f_s = 2f_a$). If we join the samples up with straight lines we produce a triangular wave. A triangular wave has frequency components at $f_a$, $3f_a$, $5f_a$, etc., so if we used a low-pass filter whose cut-off frequency was between $f_a$ and $3f_a$ we could very easily recover the original signal.

Using the same method on the samples shown in Figure 3.8c ($f_s < 2f_a$), we can see that if we join the samples together, the waveform has not retained any of the characteristics of the original sine wave. Indeed, our samples now appear to represent a sine wave with an entirely different frequency. It is no longer possible to recover the original sine wave from our samples. This is the origin of the term aliasing. The higher frequency sine wave now has an "alias" in the lower frequency sine wave inferred from our samples. Put simply, our

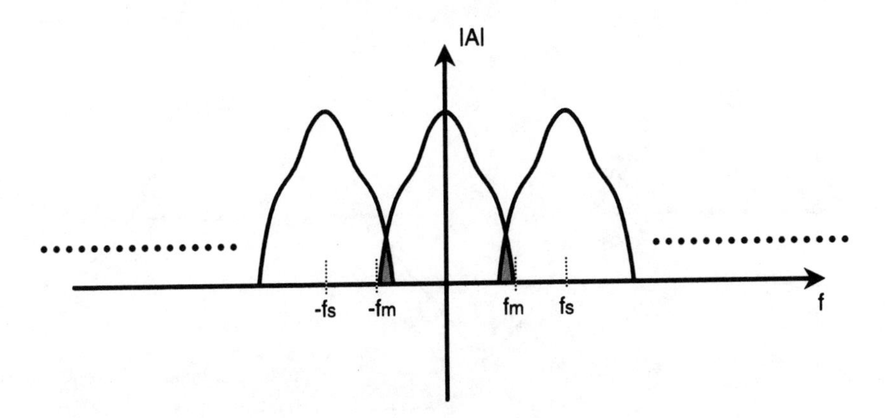


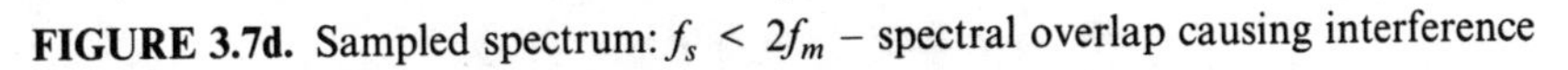

**FIGURE 3.7d.** Sampled spectrum: $f_s < 2f_m$ – spectral overlap causing interference

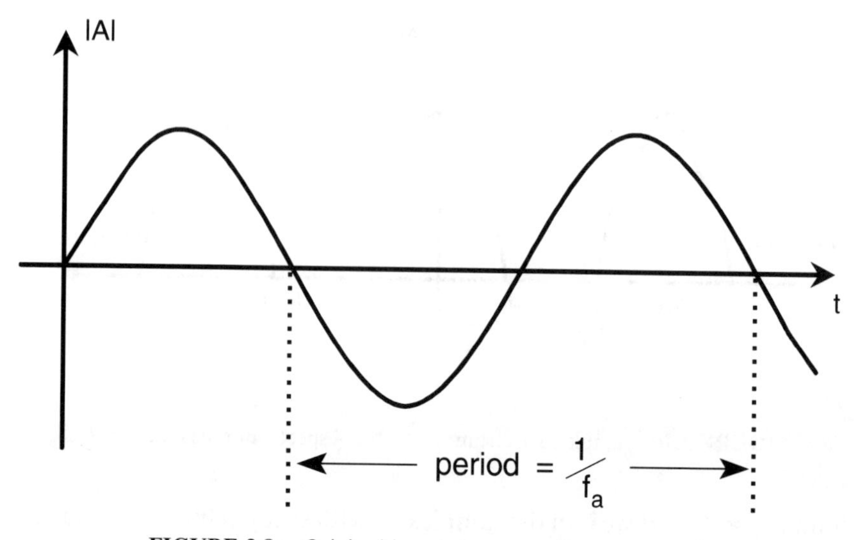


FIGURE 3.8a. Original input sine wave, frequency $= f_a$

digital samples are not representative of the input signal and therefore any subsequent numerical processing will be invalid.

The further we take the sampling frequency $f_s$ above $f_a$, the larger the component of the original sine wave in the sampled spectrum (Figure 3.8d) and the more accurate the representation of the input signal. This makes it much easier to recover the original signal. There is always a trade-off though with sampling at a high frequency: The

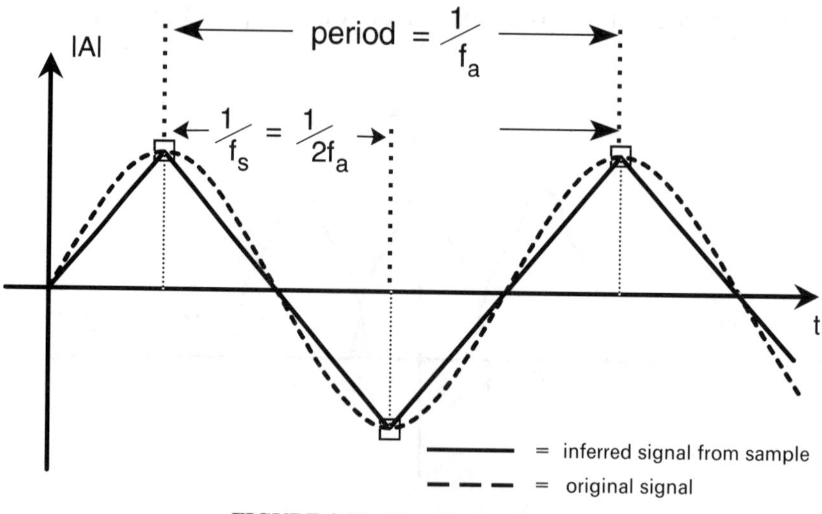


FIGURE 3.8b. Sampling at $f_s = 2f_a$

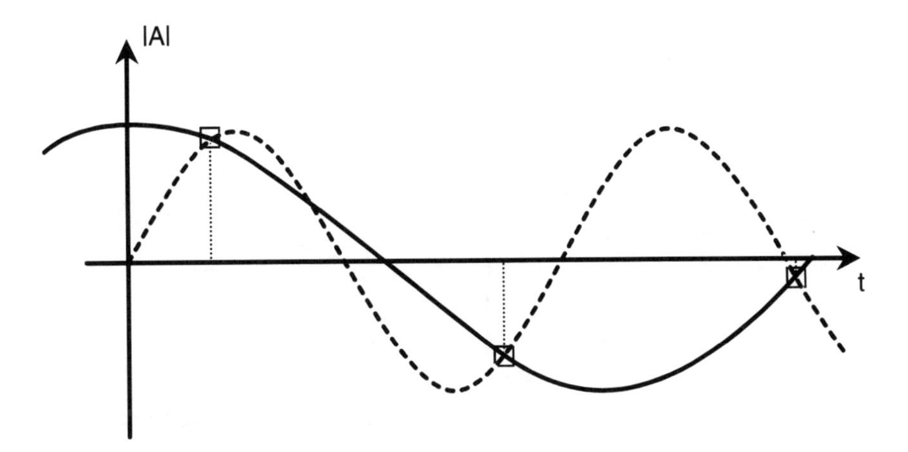


**FIGURE 3.8c.** Sampling at $f_s < 2f_a$ – aliasing occurs

DSP must calculate an output for each input sample; if we increase the sampling frequency then we reduce the time available for the DSP to perform whatever function we require.

In the real world we rarely come across a pure sine wave as shown in Figure 3.8. Typical signals have a wide spectrum, as shown in Figure 3.9. Information above a certain frequency or below a certain amplitude is usually considered surplus to requirements. For example, hi-fi manufacturers often assume that 20kHz is the highest frequency that the ear can detect. Assuming that $f_m$ is the maximum frequency of interest in a signal, we would then like to sample at the Nyquist rate, i.e., at $f_s = 2f_m$.

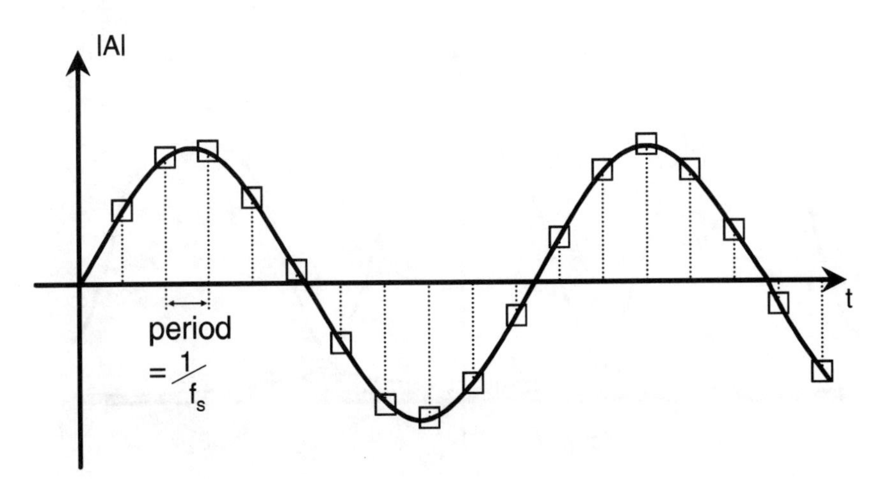


**FIGURE 3.8d.** Sampling at $f_s >> 2f_a$ – gives a more accurate representation

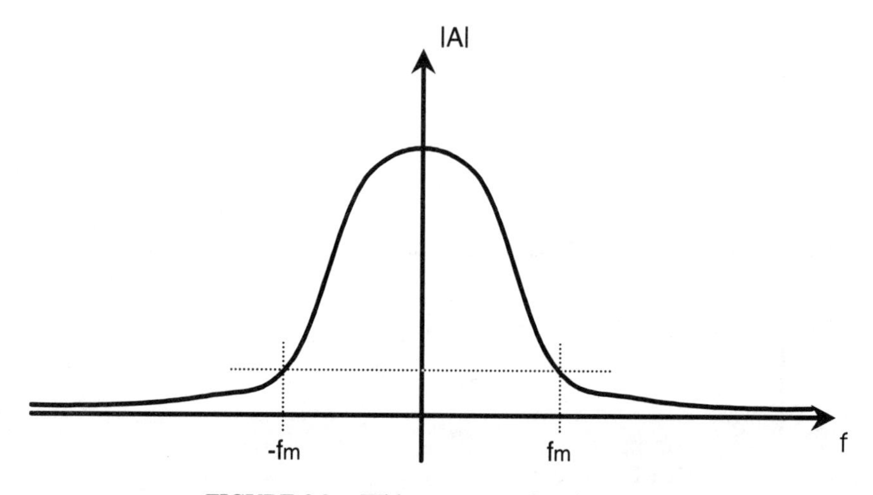


FIGURE 3.9. Wide spectrum of typical signal

If we go ahead and sample at $2f_m$ we obtain a spectrum as shown in Figure 3.10. The bands overlap and we produce aliasing in the output signal. To avoid this we must first low-pass filter the input signal to remove any frequency components above $f_m$ (Figure 3.11).

This low-pass filter is called an anti-aliasing filter. These filters are found at the input to most applications. As they are needed before the signal is sampled, they are always analog filters. It is theoretically possible to use the DSP device to remove the aliasing, but this would require a more complicated function. In practice, this is rarely done as

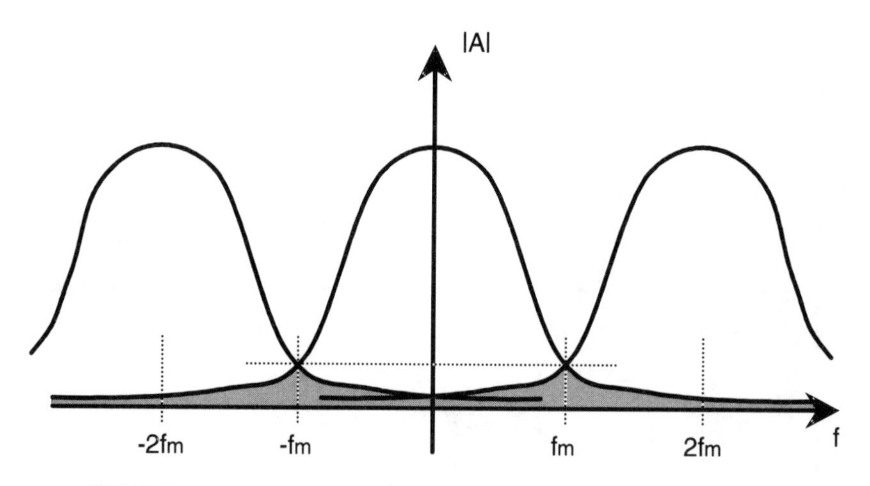


FIGURE 3.10. Sampling a wide spectrum at $f_s = 2f_m$ – aliasing occurs

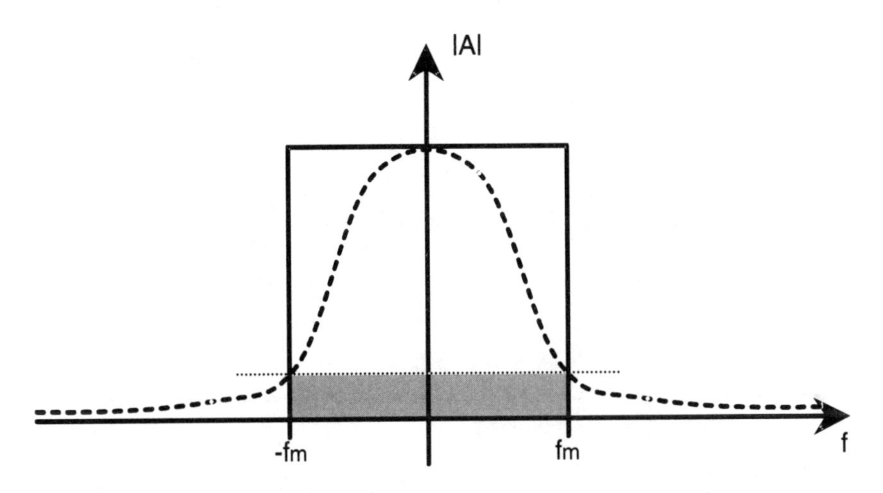


**FIGURE 3.11.** Low pass filter applied to restrict maximum frequency to $f_m$

it would absorb too much of the processing power of the DSP, which we need to perform other tasks.

The ideal anti-aliasing filter characteristic is shown in Figure 3.11. It is flat over the pass-band and zero at any other point. An ideal filter would also have a linear phase response. Variations from this ideal may have a significant effect on system performance. For example, in control applications, a badly designed filter may mean that the system spends all its time trying to compensate for the filter rather than trying to maintain a constant motor speed. In audio applications a poor phase response can lead to harmonic distortion and the degradation is often audible. For audio applications, commercially available switched capacitor filters are easy to use and provide a reasonably linear phase response.

## MATHEMATICAL REPRESENTATION OF SAMPLING

Now that we understand what sampling is actually doing to our original analog signal, we need to look a little at the mathematics that are used to describe the process. In all DSP textbooks you will find sampling described as a multiplication of the input analog waveform with a periodic delta, Dirac, or impulse function.

The impulse signal is shown in Figure 3.12. It is useful to think of the impulse function as an idealized rectangular pulse whose width is zero and amplitude infinity. The other part of the definition of the

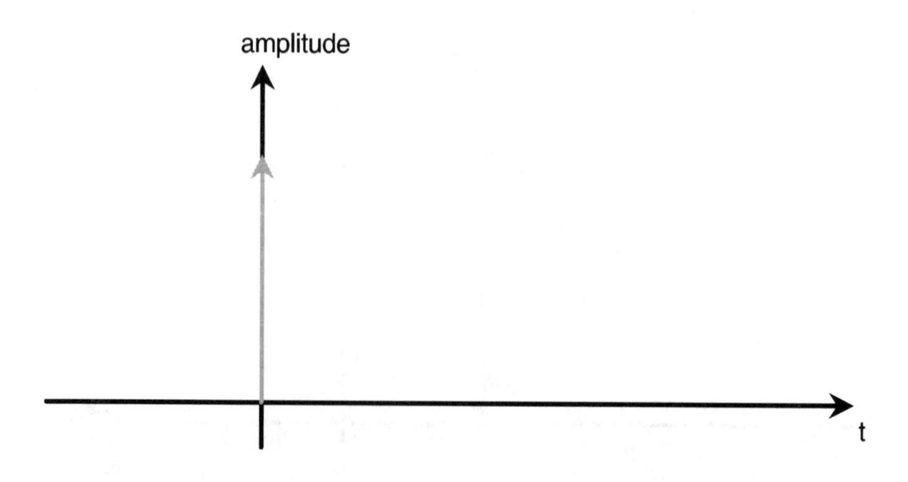


FIGURE 3.12. The impulse signal

impulse function is that the area under the "rectangle" is equal to one. This is expressed in the following formula:

$$\int_{-\infty}^{\infty} \delta(t) \cdot dt = 1$$

*Note*: This is a general formula that allows us to calculate the area underneath any signal. Take the rectangular pulse of Figure 3.13 as an example:

$$\begin{aligned}
\int_{-\infty}^{\infty} \text{pulse}(t) \cdot dt &= \int_{t=3}^{t=5} \text{pulse}(t) \cdot dt \\
\text{As the pulse} &= 0 \text{ for all the rest of the time} \\
&= \int_{t=3}^{t=5} 3 \cdot dt \\
&= [3t]_3^5 \\
&= 6 \\
&= \text{the area under the pulse}
\end{aligned}$$

We shall use "weighted" impulse functions many times throughout this book to help explain DSP techniques. A weighted impulse function is defined as:

$$\int_{-\infty}^{\infty} A\delta(t) \cdot dt = A$$

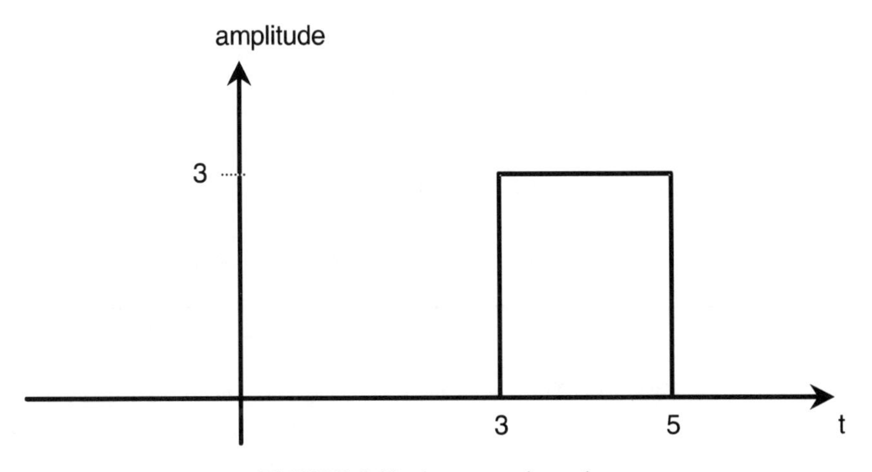


**FIGURE 3.13.** A rectangular pulse

That is, the impulse function has an area of A while the amplitude is still infinity. Nevertheless, in all the diagrams we shall draw weighted impulse functions with amplitudes proportional to the areas to allow easier interpretation. Again, this is a common technique used in DSP books.

In an idealized system our sampling waveform would consist of a train of impulse functions spaced evenly by a period $t_s$ (Figure 3.14). We can describe our idealized sampling function, $s(t)$, simply as the sum of all the individual impulse functions:

$$s(t) = \delta(t-\infty) + \ldots\ldots + \delta(t-2t_s) + \delta(t-t_s) + \delta(t) + \delta(t+t_s) + \delta(t+2t_s) + \ldots\ldots + \delta(t+\infty) \text{ or}$$

$$s(t) = \sum_{n=\infty}^{n=-\infty} \delta(t-nt_s)$$

If we then multiply these by our analog input signal, $f(t)$ (Figure 3.15), we obtain a train of pulses whose areas are equal to the amplitude of $f(t)$ at that moment in time (Figure 3.16). This is analogous to the intuitive sampling we arrived at in Figure 3.8. Mathematically, the output sampled waveform, $y(t)$, is just the multiplication of $s(t)$ with the input analog signal $f(t)$:

$$y(t) = \sum_{n=-\infty}^{n=\infty} f(t) \cdot \delta(t-nt_s)$$

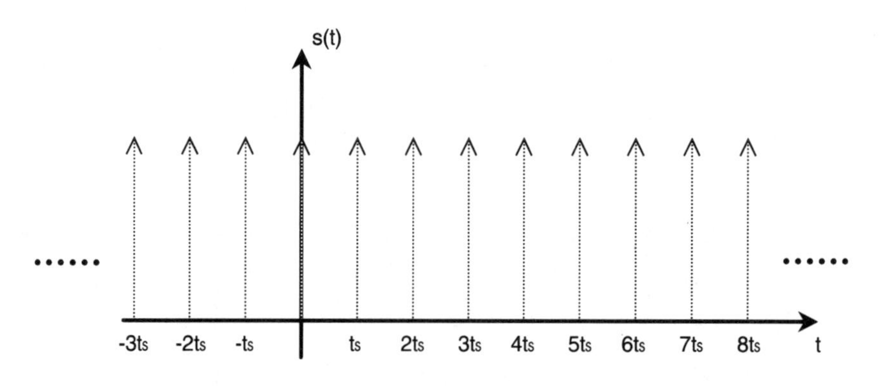


**FIGURE 3.14.** Train of evenly spaced impulse functions

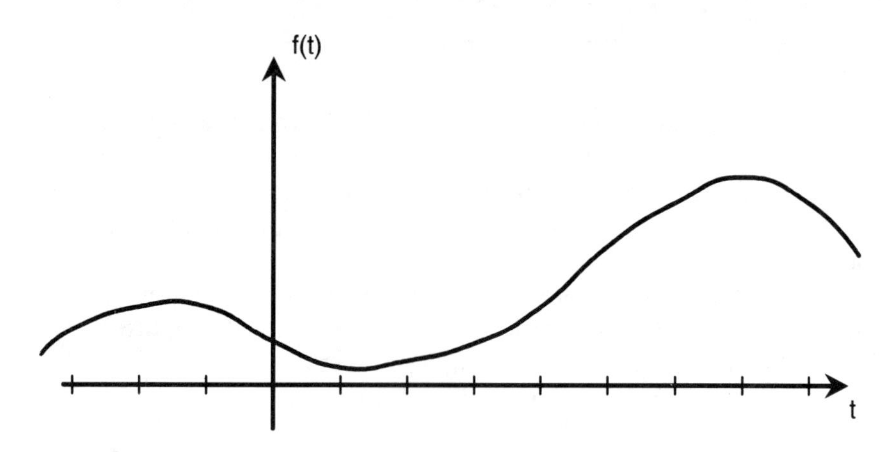


**FIGURE 3.15.** Analog input signal $f(t)$

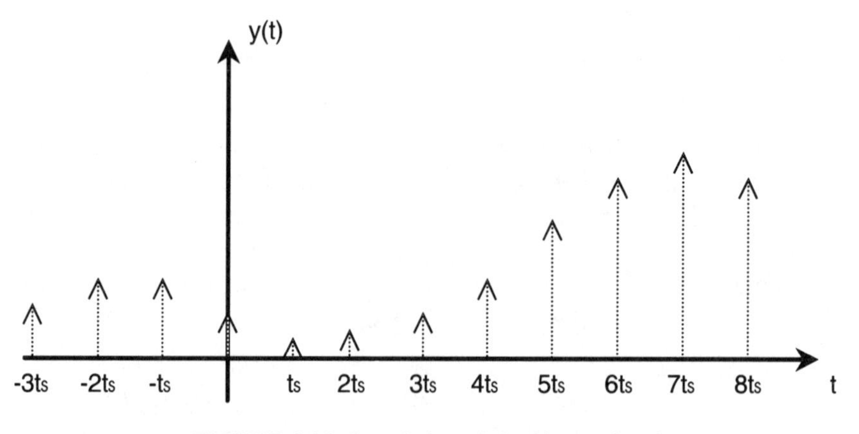


**FIGURE 3.16.** Sampled version of input signal

Great, but where does that get us? We have labored this point to introduce the weighted impulse function that we shall use in later chapters to help explain digital filtering and other techniques. This is also a starting point for deriving Shannon's theorem, but we shall not pursue that. It is enough for us to be able to understand and visualize the effects of sampling.

## DEVELOPING THE DIGITAL INPUT SIGNAL

We have now covered the first two steps in the conversion of the analog input signal to a format that can be interpreted by our digital signal processor: the anti-aliasing filter and sampling. In real-world designs, the basic sampling function shown in Figure 3.16 is replaced by a "sample and hold" circuit, which maintains the sampled level until the next sample is taken (Figure 3.17). This results in a staircase waveform, as shown in Figure 3.18.

We must now devise a method of representing these sampled values of our analog waveform as a number that can be used by our DSP. This function is called *quantizing*. Quantization is performed by an analog-to-digital converter (ADC).

Quantization can be thought of as classifying the signal into certain bands. For example, if we have just two groups we can choose a level, "a" in Figure 3.18, that we shall use as our decision point. All the samples that appear above the line will fall in group 1, and all those below the line will fall into group 0. If we look at the sequence over time we can see we have a digital stream of binary 1s

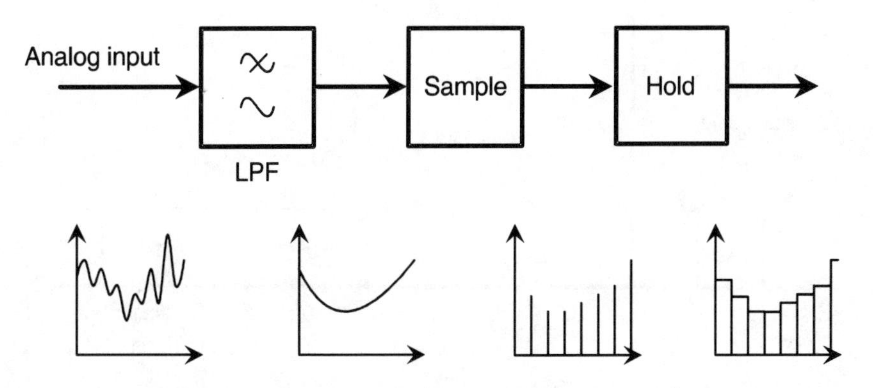


**FIGURE 3.17.** The anti-aliasing, sample and hold functions

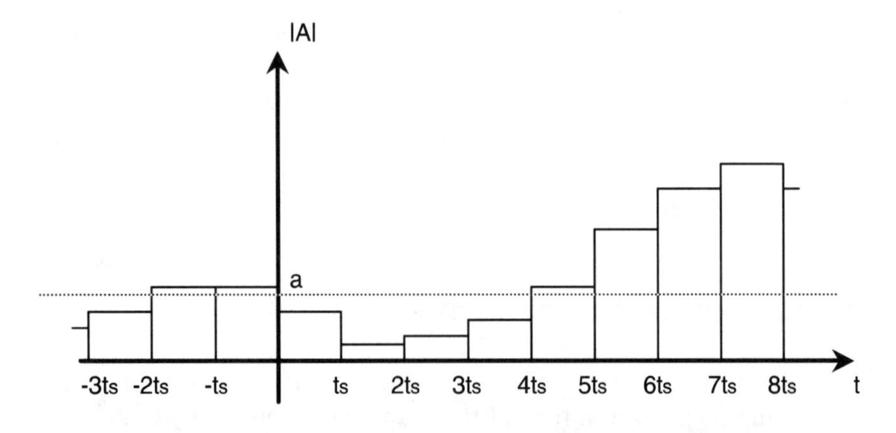


**FIGURE 3.18.** Staircase waveform resulting from sample and hold, with decision level "a"

and 0s (Figure 3.19). We have quantized the sampled input signal and can now express the data contained in it using one bit.

Taking the example further, we can use two bits to define four levels, equally spaced (00, 01, 10 and 11) (Figure 3.20), and classify the signal into these four groups (Figure 3.21). Obviously, the more levels we have, the more accurately we can describe the analog signal. In general, a DSP system will use an ADC of 10 or 12 bits. That means that the input signal will be measured against $2^{10}$ (1024) or $2^{12}$ (4096) levels. Thus, if we had an input signal that varied between 0 and 5V, the least significant bit (LSB), i.e., a single bit, would correspond to

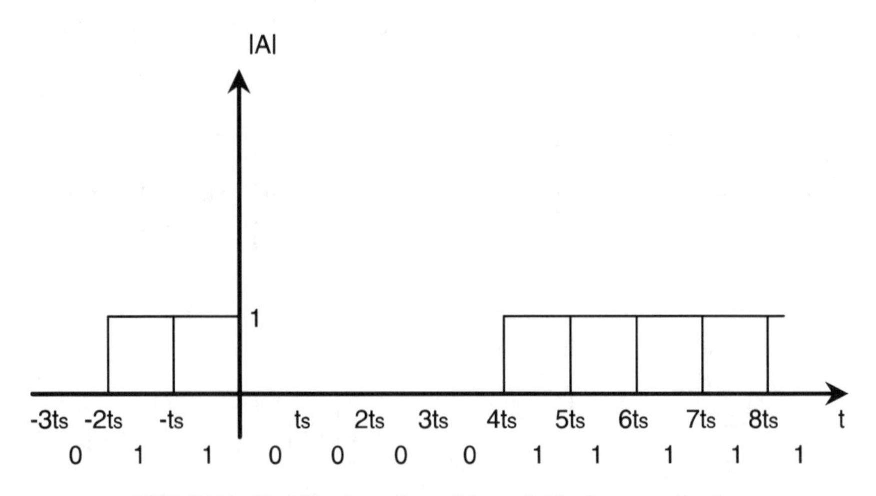


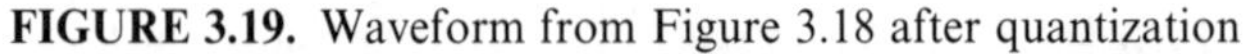
**FIGURE 3.19.** Waveform from Figure 3.18 after quantization

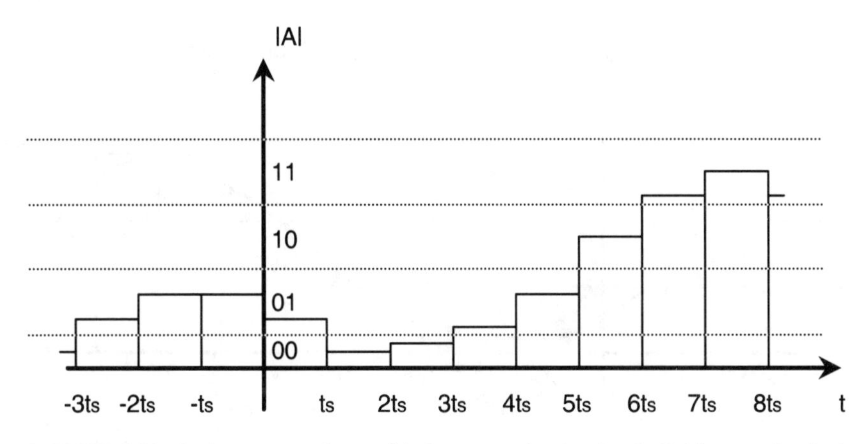


**FIGURE 3.20.** Staircase waveform with four quantization levels (2-bit quantization)

just 4.88 mV for the 10-bit ADC and 1.22 mV for the 12-bit ADC, assuming a uniform quantization step.

If we compare the quantized levels with the original waveform, we can determine the error that we have introduced (Figure 3.22). This error produces an effect called quantization noise. In audio or speech applications this error appears as noise at the output, hence the name. If we take a typical DSP application, with a 10- or 12-bit ADC, the level of quantization noise is usually negligible when compared with other noise sources in the system.

In the example above we looked at uniform quantization (Figure 3.23), but anyone delving deeper into speech or audio applications will come across a bewildering array of different quantization schemes. In

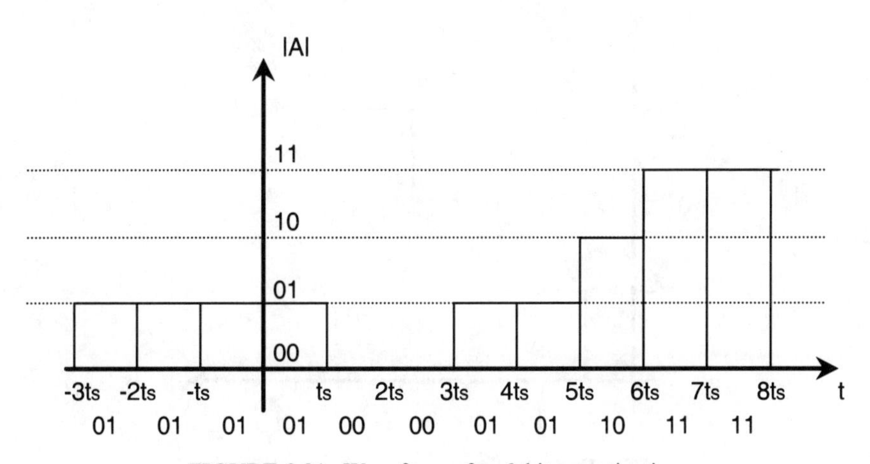


**FIGURE 3.21.** Waveform after 2-bit quantization

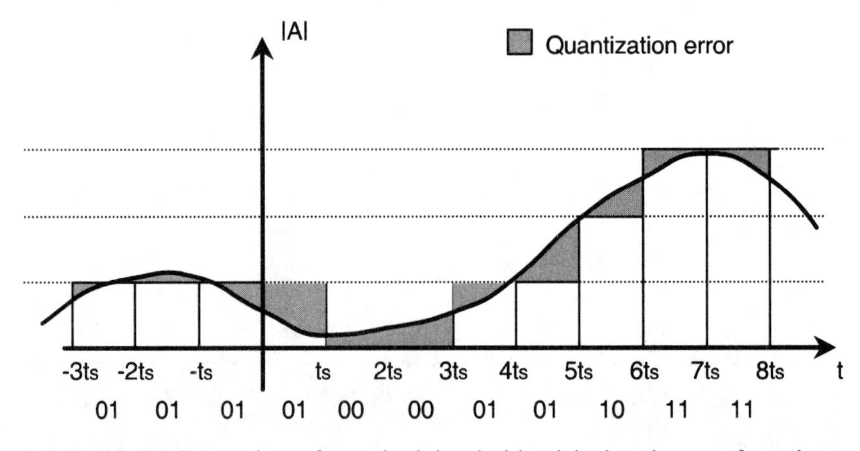


**FIGURE 3.22.** Comparison of quantized signal with original analog waveform shows quantization error

speech there are loud sounds, such as vowels, and much softer sounds, such as consonants. A typical speech waveform is shown in Figure 3.24. The large amplitude components could be the "oh", or "eh" sounds, and the smaller amplitudes may correspond to the "sh" or "ph" sounds.

A uniform quantization scheme, as shown in Figure 3.24, would ensure that we could adequately represent the loud sounds, but most of the softer sounds would be given the same binary value. This means that we would not be able to distinguish between the softer sounds, which will greatly impair the quality of the subsequent speech output.

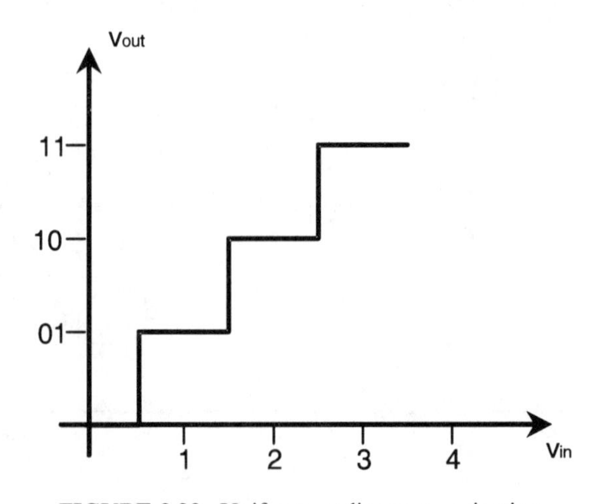


**FIGURE 3.23.** Uniform, or linear quantization

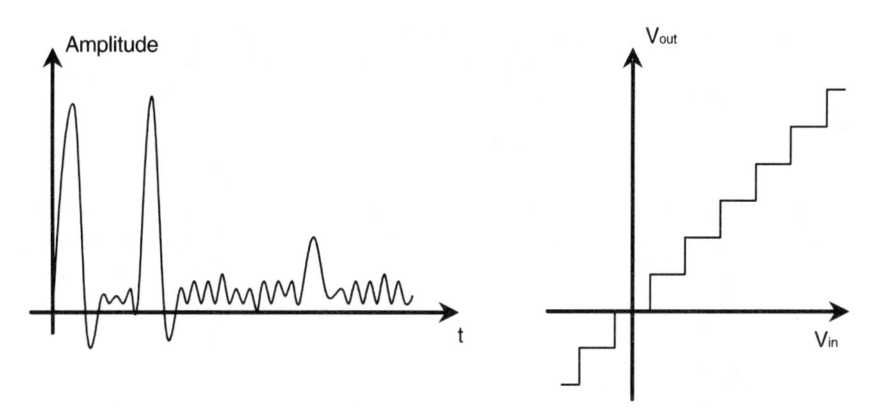


**FIGURE 3.24.** A typical speech signal and uniform quantization scheme

To get around this problem we adopt a different quantization scheme whose step size varies according to the signal amplitude. In the case of our speech signal we would ensure that there are more levels at the lower amplitudes (Figure 3.25).

In practice, the quantizer has a uniform step size and it is the input signal that is compressed. In our speech example, the higher amplitude sounds will be compressed leaving the lower amplitude sounds unchanged. The overall effect is identical to nonuniform quantization. After processing, the signal is reconstructed at the output by expanding it. This process of expansion and compression is called *companding* (COMpressing and exPANDING).

The most widely used application of companding is the public

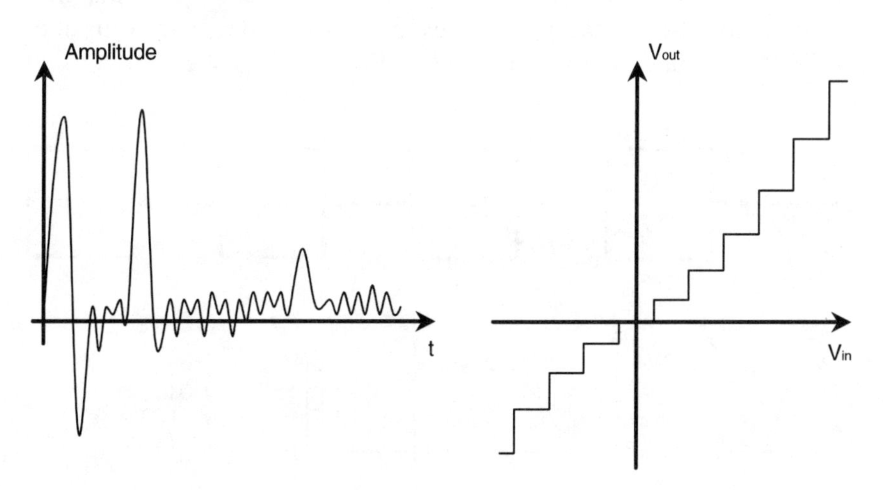


**FIGURE 3.25.** A typical speech signal and non-uniform quantization scheme

telephone network. There are two distinct companding schemes used in Europe and the United States. The method used in Europe is called A-law, and the U.S. method is called $\mu$-law. Definitions of the two schemes can be found in the references at the end of this chapter.

In speech coding, there are many other types of quantization used, depending on the nature of the expected input signal. There are also adaptive quantization schemes, which we shall investigate later in Chapter 6.

## REAL-WORLD ADCS

We have now covered all the functions necessary to convert an analog signal into a signal that our DSP can understand (Figure 3.26). The noisy analog input signal is first low-pass filtered to remove any frequencies above $0.5f_s$, then passed through a sample and hold and the resulting levels converted to binary values. The digital bit pattern output from the ADC is then passed to the DSP.

Let's now look at what ADCs are commercially available. There are many different types in today's market, most of which also contain the sample and hold circuit. We shall discuss the four most popular types of ADC, each of which uses a different method of quantization.

### Successive Approximation ADC

The successive approximation ADC produces an $n$-bit output in $n$ cycles of its clock by comparing the input waveform with the output of a digital-to-analog converter (DAC) (Figure 3.27). We shall look at

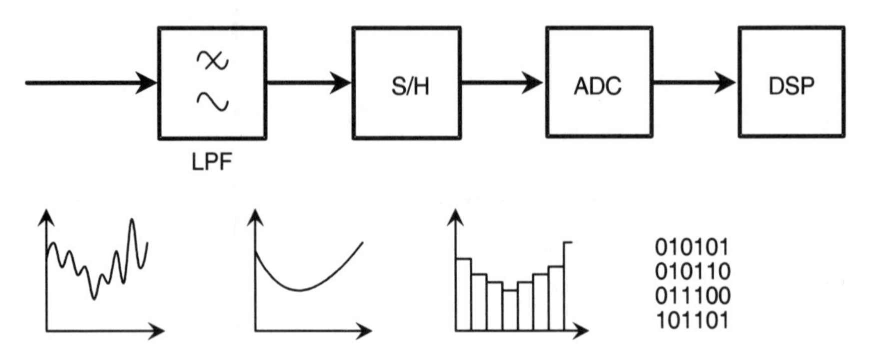

FIGURE 3.26. The complete analog to digital conversion process

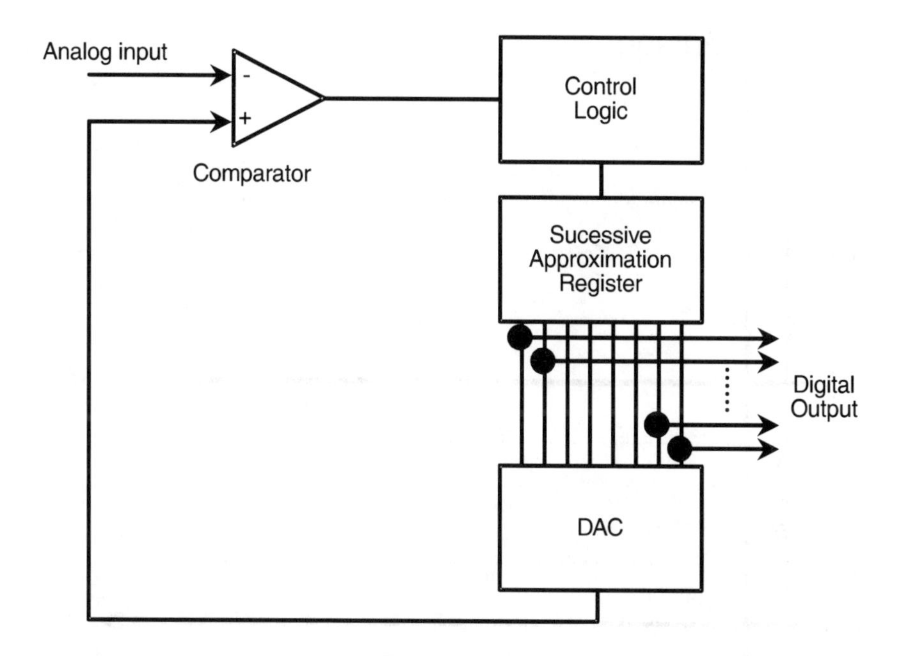


**FIGURE 3.27.** Successive approximation ADC

DACs later in this chapter. The technique used by this device is to keep splitting the voltage range in half to determine where the input signal lies.

Looking at the example in Figure 3.28, the input voltage is shown as $V_{in}$. This ADC is a 2-bit device for ease of explanation, so there are four possible ranges in which the voltage can lie. On the first cycle, the output of the DAC is set to half full-scale, which means that the overall voltage range of the device is divided in two. The input signal is then compared with the DAC voltage and because it is higher than the output of the DAC, the most significant bit (MSB) is set to a 1.

This then causes the DAC output to change to three-quarters full-scale, and $V_{in}$ is compared with this new value. $V_{in}$ is now below the voltage output of the DAC so the next bit is set to 0. Hence, in two cycles we have determined that $V_{in}$ should be classified as 10. Obviously, this technique can be extended to many bits.

Successive approximation ADCs are relatively cheap to produce and are generally accurate and fast. Conversion times vary from $< 1\mu s$ to $50\mu s$ with between 8- and 12-bit accuracy. This type of converter operates on only a brief sample of the input waveform, and spikes in the input can prove disastrous. Also, its ability to follow changes in

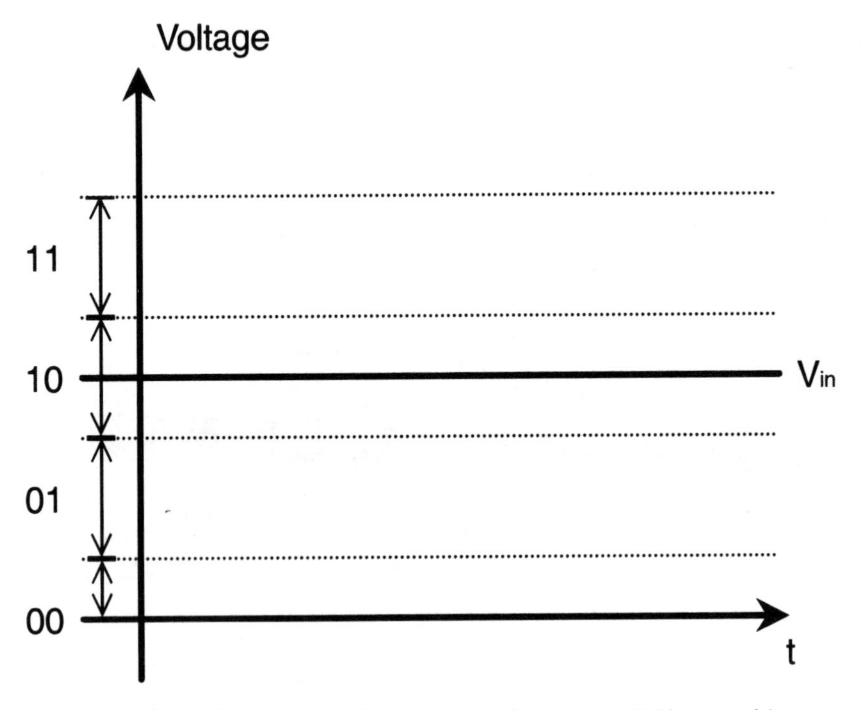


**FIGURE 3.28.** The successive approximation process (2-bit example)

the input signal is limited by its internal clocking rate, so it may be slow to respond to sudden jumps in the input signal.

### Dual Slope ADC

The dual slope ADC uses a capacitor connected to a reference voltage. The capacitor voltage starts at zero and it is charged for a set time, $t_1$ by the output voltage from the sample and hold, $V_{in}$ (Figure 3.29). The capacitor is then switched to a known negative reference voltage and charged in the opposite direction until it reaches zero volts again. The time taken for this second charge cycle is recorded using a digital counter. With the counter set initially to zero, the final counter value gives a digital output proportional to $V_{in}$.

This technique is very precise and can produce ADCs with high resolution. The dual slope technique ensures that most component variations are cancelled out. For example, drifts or scale errors have no effect as we start and finish at the same voltage. As both charging rates are inversely proportional to the value of capacitor used, temperature, tolerance and ageing also have little or no effect.

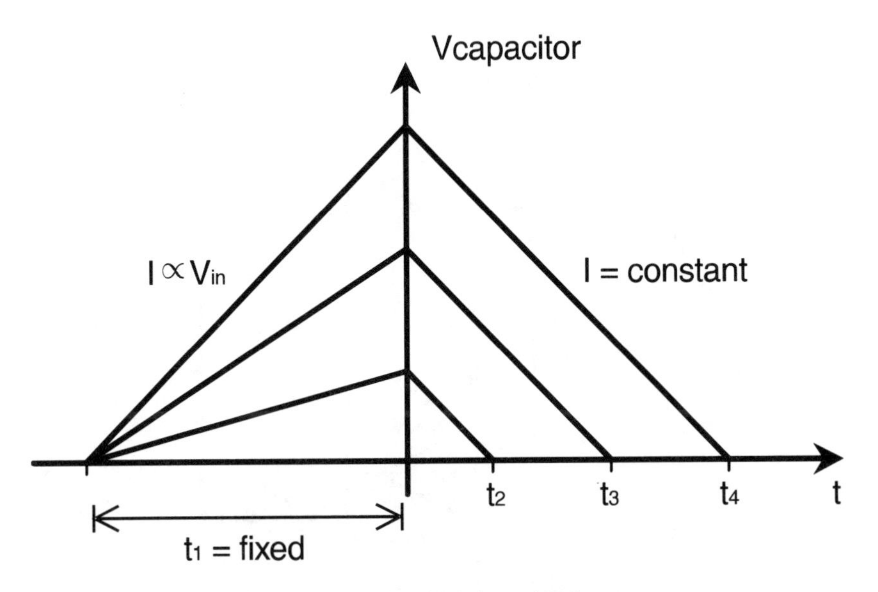


**FIGURE 3.29.** The dual slope ADC process

The disadvantages of these converters are that they are very slow and generally more expensive than the successive approximation ADCs.

### Flash ADC

Parallel flash ADCs convert the analog input voltage faster than other types of ADCs. Figure 3.30 illustrates the principle with a 2-bit ADC. The input voltage, $V_{in}$, is compared with a set of reference voltages developed across a ladder of equally valued resistors. So, for an $n$-bit ADC we require $2^n$ resistors. The tapped voltages are then passed to comparators and the digital output decoded. The resistors must be matched and laser trimmed for accuracy; hence commercial flash ADCs are not usually available above 8-bit accuracy (256 matched resistors).

### Sigma Delta ADC

The sigma delta modulator was first introduced in 1962, but until recent developments in digital VLSI technologies it was difficult to manufacture cost-effectively. One of the major advantages of the

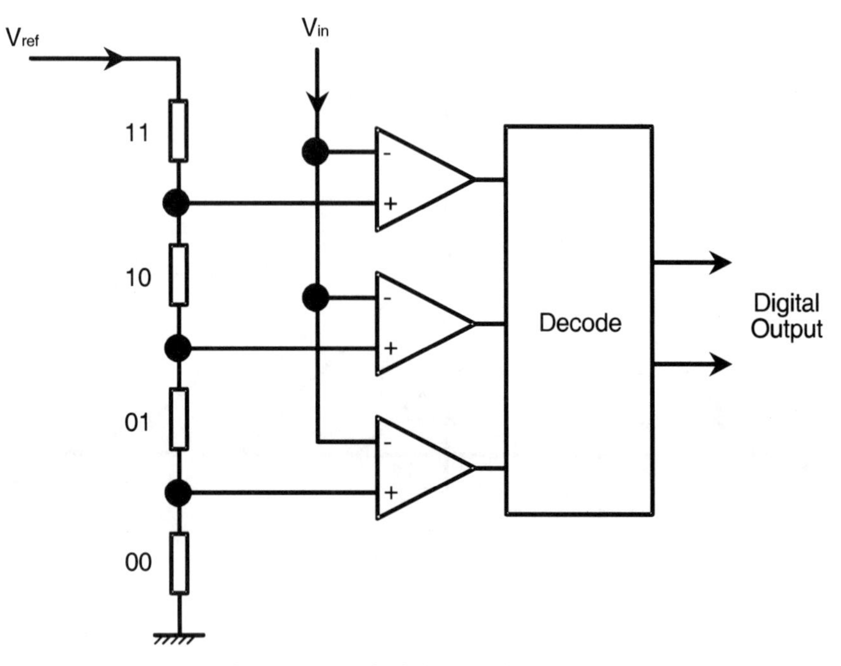


**FIGURE 3.30.** Flash ADC (2-bit example)

sigma delta technique is that it is able to use digital filtering and approximately 90% of the silicon die area is purely digital. As a result, sigma delta converters are now available with high resolution and good noise characteristics at competitive prices.

The sigma delta ADC also benefits from all the usual digital advantages, i.e., higher reliability, higher stability and increased functionality. In addition, as the sigma delta converter uses mainly digital techniques, it is possible for it to be integrated onto the same silicon die as the DSP. This reduces the number of integrated circuits in the system and hence the overall cost, while increasing reliability.

Successive approximation, dual slope and flash analog-to-digital converters are all based on the principle of sampling at around the Nyquist frequency and require a good anti-aliasing filter. They achieve their high resolution using precise component matching and laser trimming. Sigma delta ADCs, on the other hand, use a low-resolution ADC (one-bit quantizer) with a sampling rate many times higher than the Nyquist frequency. This is followed by "decimation" in the digital domain, which lowers the output frequency and increases the accuracy.

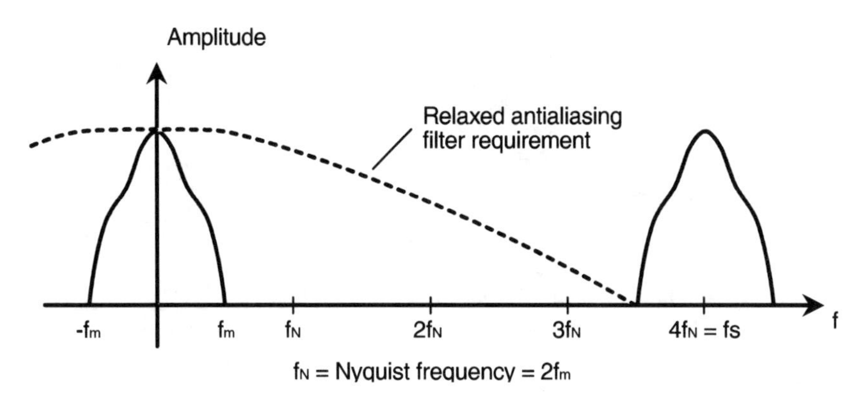


**FIGURE 3.31.** Sampling at four times the Nyquist frequency

Let us sample the input waveform at four times the Nyquist frequency and investigate the effects (Figure 3.31). We can see that the requirements for our anti-aliasing filter are significantly relaxed. The relative area over which we require the response to be flat becomes smaller, the roll-off rate of the filter is lower, and the filter becomes much easier to design. Look back to Figure 3.11 for the "perfect" anti-aliasing filter.

Sampling at a higher rate, or oversampling, also has another benefit. In the process of quantization, the resulting noise power is spread evenly over the complete spectrum. With the sigma delta ADC we now have the same total noise power spread over a larger frequency range. The noise power within the band of interest ($-f_m$ to $+f_m$ in Figure 3.32) is therefore lower.

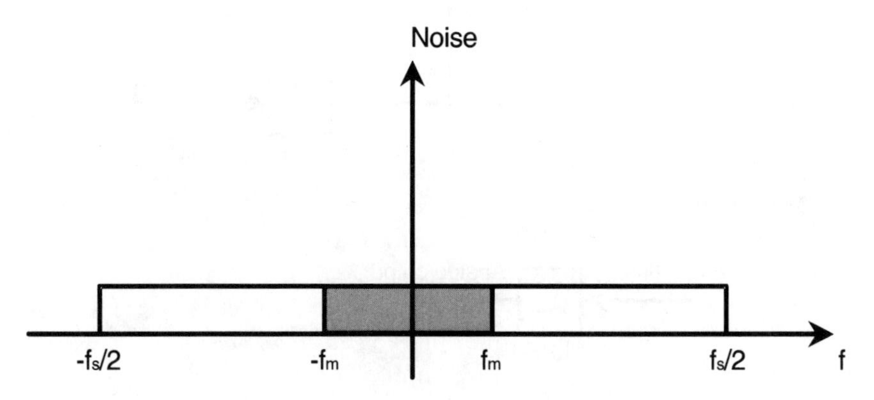


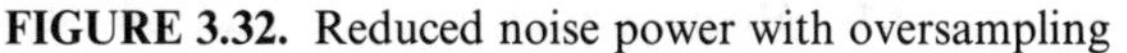
**FIGURE 3.32.** Reduced noise power with oversampling

(a) Modulator

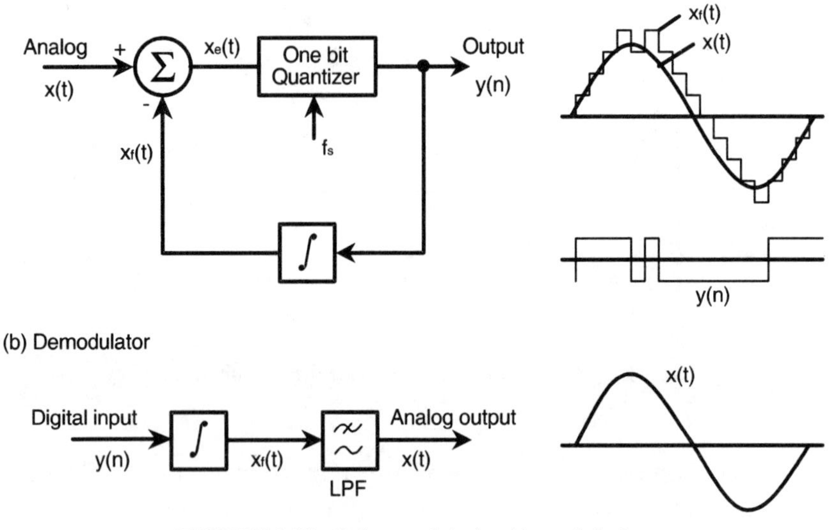


**FIGURE 3.33.** Delta modulation/demodulation

The one-bit quantization of the sigma delta ADC uses a method that was derived as an extension to a modulation technique called delta modulation. This is based on quantizing the difference between successive samples rather than the absolute value of a sample

(a) Modulator

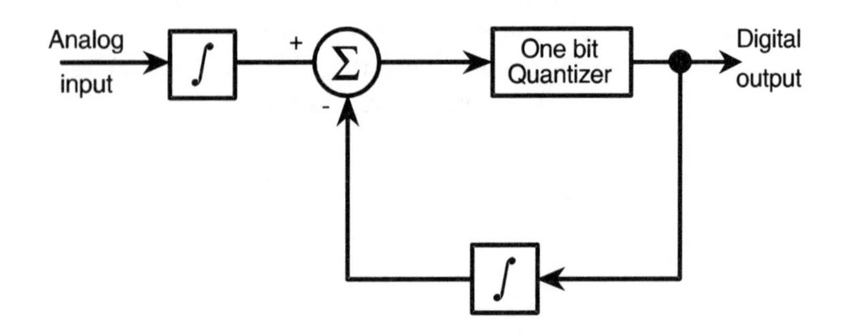


(b) Demodulator

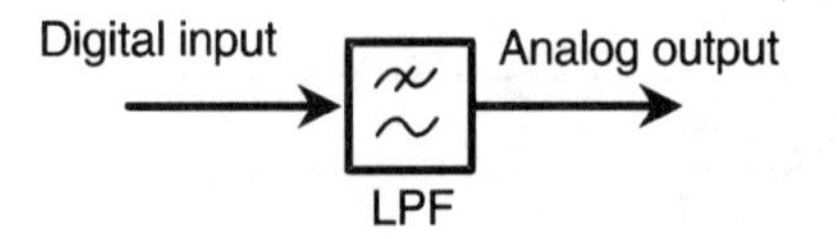


**FIGURE 3.34.** Modified delta modulation/demodulation

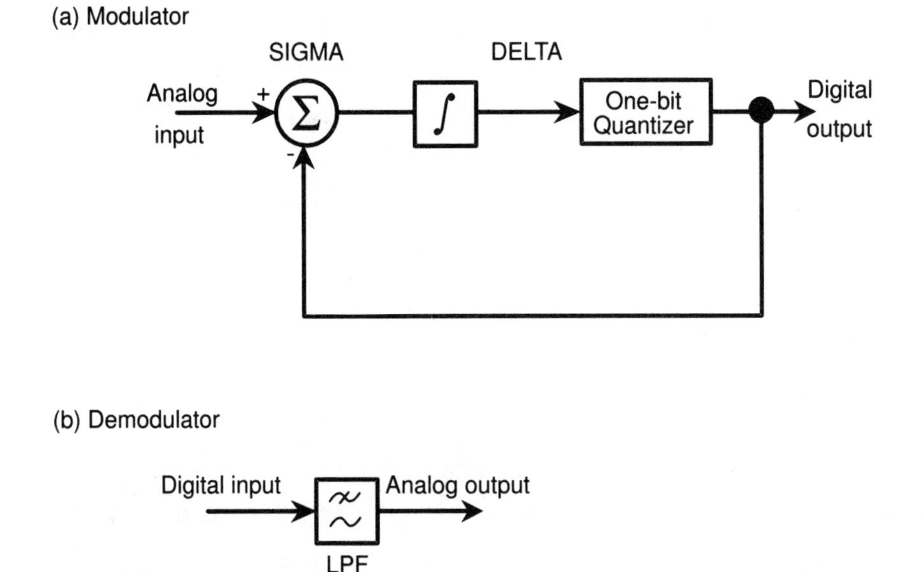


**FIGURE 3.35.** Sigma delta modulation with combined integrators

(Figure 3.33). Sigma delta modulation is achieved by moving the second (demodulation) integrator into the input stage (Figure 3.34). As integration is a linear function, this can be done without affecting the operation of the system. Furthermore, the two integrators can be combined, as shown in Figure 3.35.

The name sigma delta is derived from the summation point (sigma) followed by the delta modulator (integrator and one-bit quantizer). The noise performance of such a coder is frequency dependent. The loop acts as a low-pass filter for the input signal and a high-pass filter for the noise introduced by the quantizer. This noise-shaping property is well suited to digital audio applications.

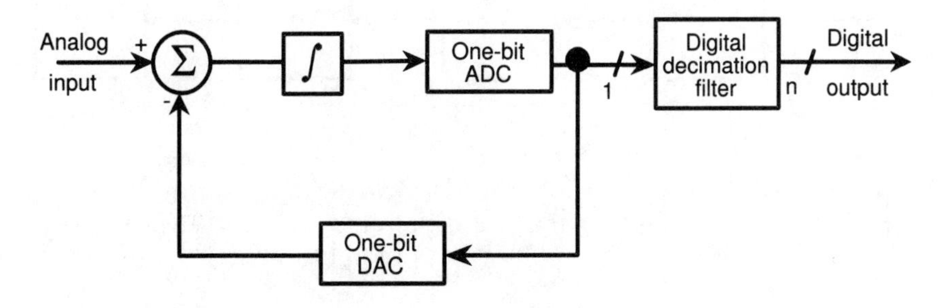


**FIGURE 3.36.** A sigma delta ADC

Output from one-bit ADC :-

0 0 1 0 1 1 0 0 1 0 1 0 1 1 1 1 1 0 1 0 1 1 1 0 1 0 1 0 1 1 1 0 1

| 4x0 3x1 | 3x0 4x1 | 2x0 5x1 | 3x0 4x1 |
|---|---|---|---|
| = 0 | = 1 | = 1 | = 1 |

Output from decimation filter (÷7)

| 0 | 1 | 1 | 1 |
|---|---|---|---|

**FIGURE 3.37.** Digital decimation filter operation

Further discussion of this property can be found in the references at the end of this chapter.

A block diagram of a sigma delta ADC is shown in Figure 3.36. The one-bit oversampled digital output is passed to a decimation, or rate reduction, filter in the digital domain, which averages the values and produces an $n$-bit result at a lower frequency (Figure 3.37). We shall see later how we can implement digital filters using a DSP (Chapter 4).

There are many other types of analog-to-digital converters available on the market today. We have discussed only the four most popular methods.

This ends our description of the anti-aliasing, sample and hold and analog-to-digital blocks of our DSP system (Figure 3.38). Now we shall turn our attention to the output side of our DSP system, i.e., the digital-to-analog converter or DAC.

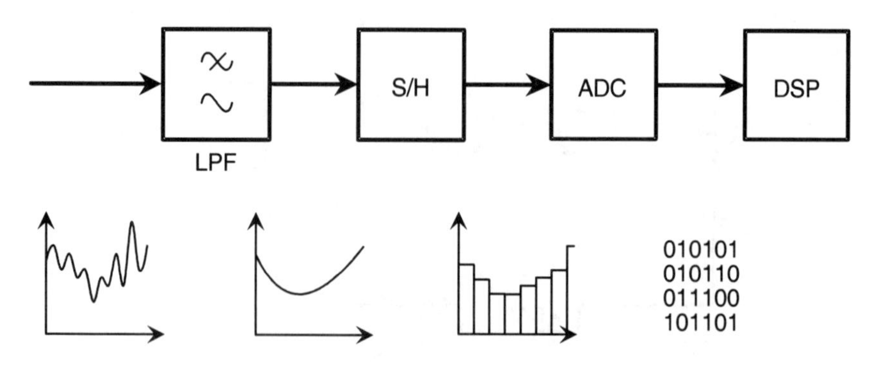


**FIGURE 3.38.** The complete analog to DSP process

## RECONSTRUCTING THE ANALOG SIGNAL FROM THE DIGITAL OUTPUT

In many DSP applications, we must reconstruct an analog signal after the digital processing stage. This function is performed using a digital-to-analog converter (DAC). The cost of the DAC is usually considerably less than the ADC used and it has settling times from around 100ns for 8-bit conversion to approximately 1.2$\mu$s for 12-bit, depending on its design.

## MULTIPLYING DAC

This is the most common form of DAC. The output is the product of an input current, or voltage and an input digital code. A current source multiplying DAC is shown in Figure 3.39. The input digital code is used to turn on a selection of current sources that are then summed. The resulting current is either brought out as is or converted to a voltage using an operational amplifier. Generally, in current source DACs the sources are always on, but switched to ground when not in use. Multiplying DACs have the advantage that they are fast, with settling times of 100ns or less.

The other multiplying technique uses a voltage source and a set of scaled resistors again into a summing junction (Figure 3.40). The state of a single bit is used to set the switch to either $V_{cc}$ or GND. This circuit generates an output proportional to the weighted sum of the input voltages.

Not all multiplying DACs have built-in references ($V_{cc}$); some

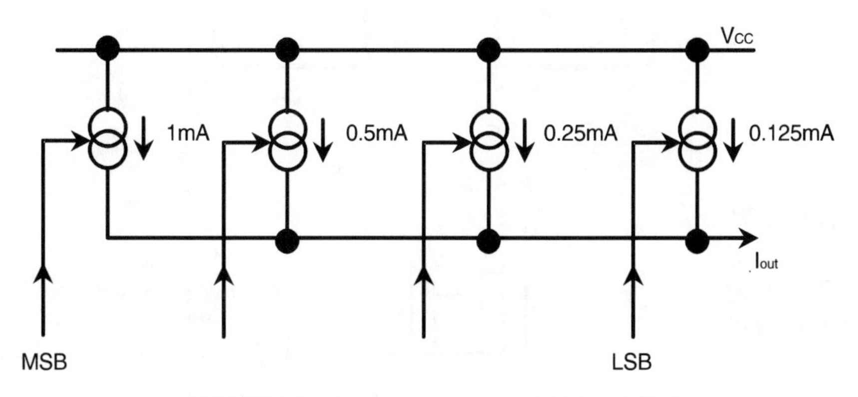


**FIGURE 3.39.** A current source multiplying DAC

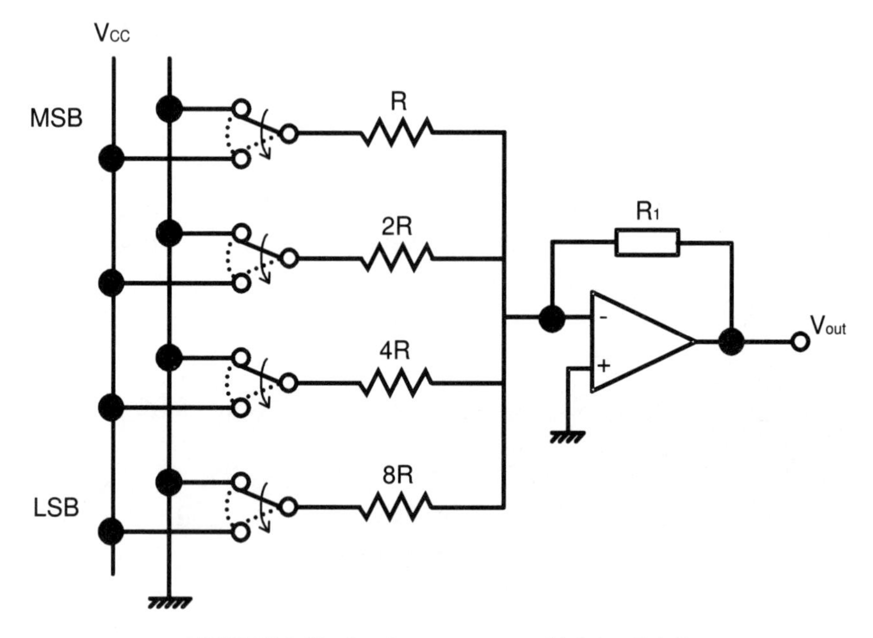


**FIGURE 3.40.** A voltage source multiplying DAC

allow the user to input his own reference voltage. This effectively enables the user to specify the accuracy of his DAC.

Most commercial DACs are Zero-Order-Hold (ZOH), which means that they convert the binary input to the analog level and then simply hold that value until the next sample (Figure 3.41)

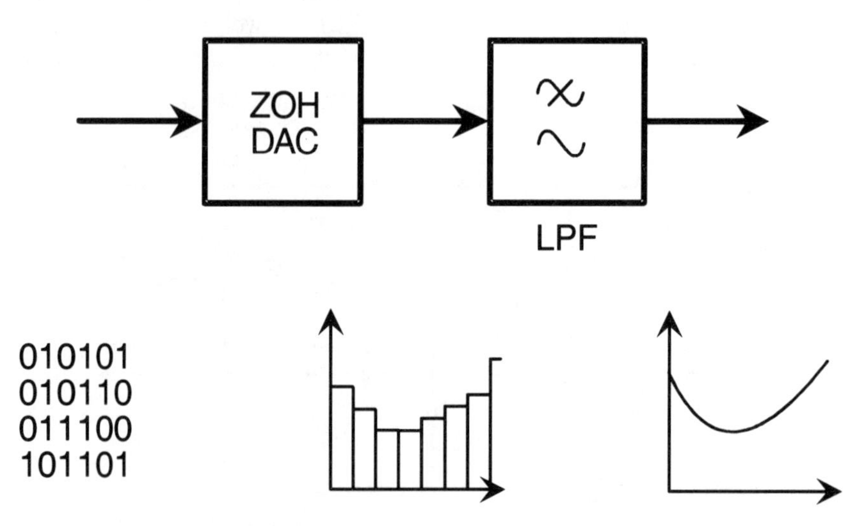


**FIGURE 3.41.** Zero order hold (ZOH) DAC output

producing a staircase waveform. A reconstruction filter is then used to smooth the output of the DAC. This final filter is also sometimes (confusingly) called an anti-aliasing filter.

## MULTIRATE AND BIT STREAM DACS

The types of DACs that we have already described bear the disadvantage that the most significant bit (MSB) must be extremely accurate and the DAC must maintain this accuracy over the device's total temperature range and over the lifetime of the device. A 16-bit DAC, for example, must have an MSB accurate to one part in 65,536 ($2^{16}$). In the previous example the output was generated from either a voltage or current source; hence the output of this source must be accurate to one part in 65,536. Controlling the output of a voltage or current source to this level of accuracy is a very difficult task. Bit stream DACs utilize a concept that we saw implemented in sigma delta ADCs: using a much higher sampling frequency in exchange for a smaller number of bits in the quantized signal.

The concepts behind bit stream DACs are fairly simple. Let us take an input signal consisting of $n$ bits at a frequency $f_{in}$. The first part of any bit stream DAC is an integrator (Figure 3.42), which in the simplest form just adds the previous value to the present value. The next step is to requantize the new value into ($n$-2) bits, i.e., truncating the last two bits of the value. The output from this is fed back and subtracted from the next input sample, which allows us to compensate for the error introduced by truncation. This operation is repeated four times ($2^2$) for every input sample. The result of all this is that we get four outputs that are quantized to ($n$-2) levels – i.e., more samples with less accuracy.

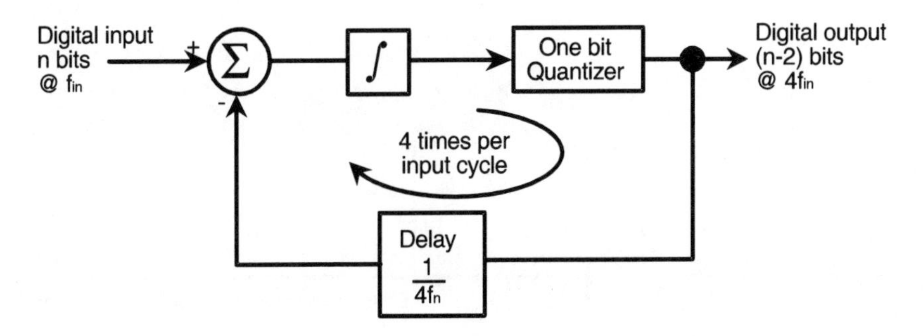


**FIGURE 3.42.** A bit stream DAC

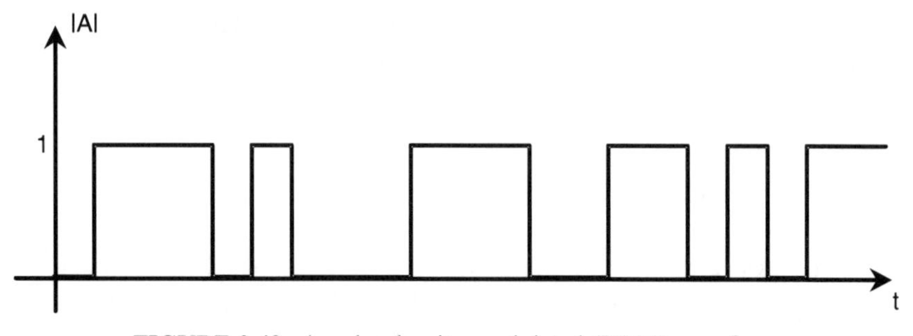


FIGURE 3.43. A pulse density modulated (PDM) waveform

This technique can obviously be extended by increasing the number of extra samples per input cycle and also by increasing the range of the integrator. The method used varies among manufacturers, who claim superior noise performance with their particular $n^{\text{th}}$-order XYZ system. References to several papers on the subject can be found at the end of this chapter.

In most systems the output of the oversampling system is a one-bit representation of the original signal, which is a pulse density modulated (PDM) waveform (Figure 3.43). This is a pulse stream that has been modulated in both frequency and width. Although this is the most favored method, some bit stream DACs produce pulse width modulated (PWM) outputs (Figure 3.44) where only the widths of the pulses are modulated. The advantage of PWM is that it is much simpler and only a low-pass filter is needed to convert the output pulse train into an analog signal. In the case of the PDM system a more sophisticated circuit must be used: A one-bit DAC is realized using a switched capacitor network followed by a simple low-pass filter to produce a smooth continuous output voltage.

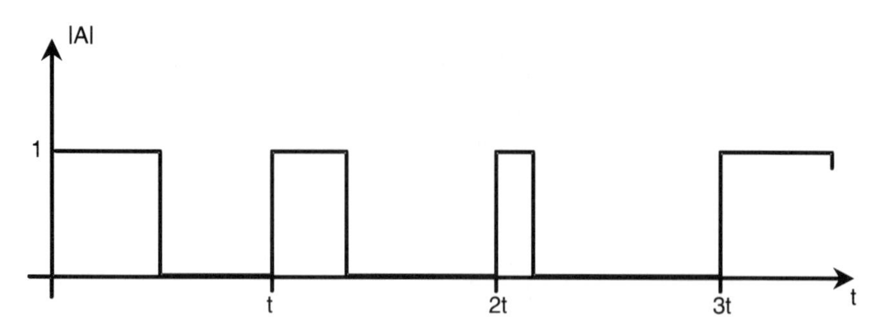


FIGURE 3.44. A pulse width modulated (PWM) waveform

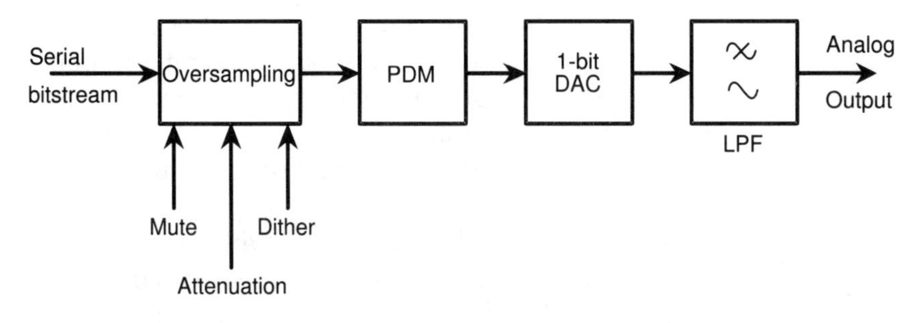


**FIGURE 3.45.** A PDM bit stream DAC

The basic block diagram of a PDM bit stream DAC is shown in Figure 3.45, showing the functions that we talked about in the previous paragraph. In addition, we have three more input signals to the oversampling block: mute, attenuation and dither. The first two signals are self-explanatory, so we shall simply examine the effects of dither here. Let us take the analog signal shown in Figure 3.46. This signal could represent the low-amplitude, high-frequency component of a musical signal. With the first quantization level as shown in the figure this signal would be quantized as a 1 for the whole of the sample period shown.

There is a significant difference between the mean of the new digital signal and the mean of our original analog signal ($V_m$). This results in an error between the two signals that appears as a "deadening" of the audio output. One way to overcome this is to have more quantization

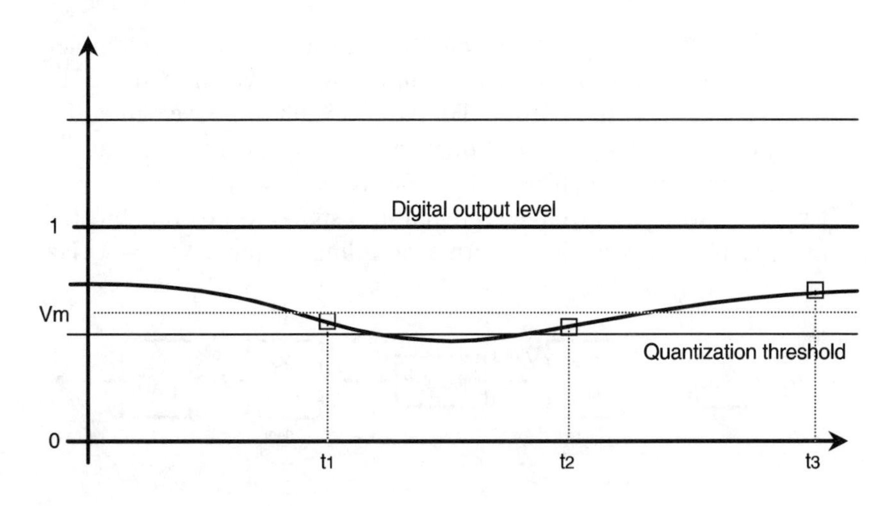


**FIGURE 3.46.** Quantization and the need for dithering

levels so that we capture more of the information, but is this really necessary? This type of signal in audio systems is very similar to noise and it can be represented by a pseudorandom digital sequence. Hence, we can re-create some of the variation below the threshold of the least significant bit by adding a pseudorandom digital sequence to the LSB of the quantized digital bit stream.

There are many alternative dither patterns, with each scheme designed to enhance a specific characteristic of the original analog waveform. However, the analysis of the effect of these schemes is far beyond the scope of this book. The reader will find dither techniques mainly referred to in conjunction with DACs, but it is clear that the technique is also applicable to ADCs.

Dither signals also have another advantage in bit stream DACs. One of the main problems of bit stream DACs is the tendency of the re-quantized signal to "hang up", or stick at an incorrect value, causing an erroneous dc component in the output signal. For example, if we are requantizing a 2-bit signal into a single bit, a long series of 2s should average out to $\frac{2}{3}$ in the output signal.

$$\begin{aligned}
2 \text{ bits} &\equiv 1 \text{ bit} \\
0 &\equiv 0 \\
1 &\equiv \tfrac{1}{3} \\
2 &\equiv \tfrac{2}{3} \\
3 &\equiv 1
\end{aligned}$$

What sometimes happens is that the output of the oversampler sticks at 1 until the next change in input signal, which leads to an erroneous dc level at the output. To avoid this hang-up we can add a pseudorandom noise signal, or dither, to the data stream such that the probability of long sequences of any one value is very low.

If we were to try and reduce a real audio signal to just one bit, for example, a 16-bit audio signal with a sampling frequency of 44.1kHz

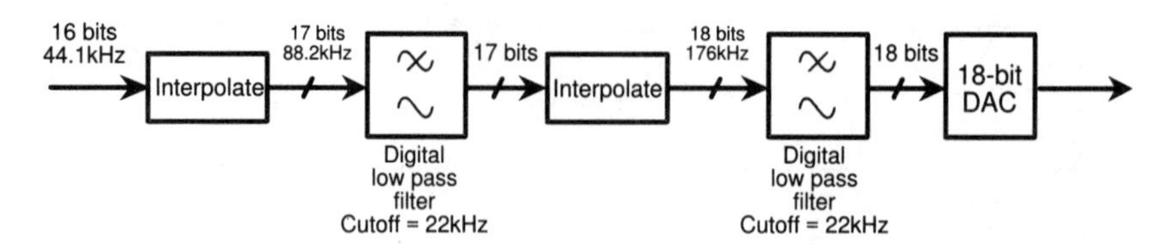


**FIGURE 3.47.** A typical CD interpolation/oversampling scheme

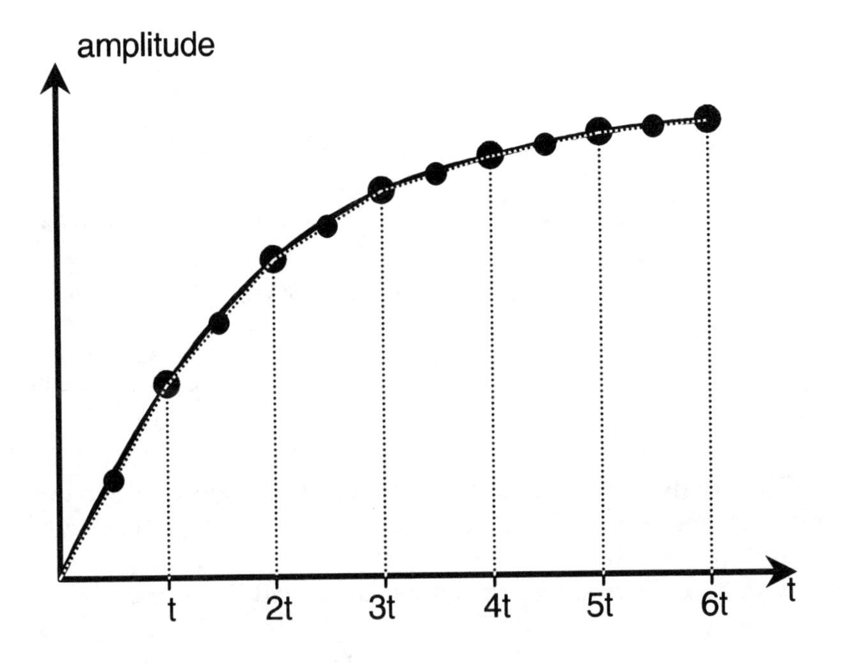


FIGURE 3.48. The interpolation process

(the sampling frequency for compact discs), we should require a frequency of $2^{16} \times 44.1 \times 10^3$, or approximately 3GHz. Even with today's advanced VLSI techniques we would find this frequency difficult to implement. Thus, the majority of bit stream DACs can be split into two categories: those that oversample to word widths of greater than one bit and those that combine pure bit stream techniques with interpolation.

Oversampling by interpolation was very popular in digital audio systems a few years ago. Figure 3.47 shows the block diagram of a typical system. By interpolation, we mean the computation of the average value half way between two consecutive samples (Figure 3.48). When we compute an average, the two original points and the new value lie on a straight line. Obviously, the analog signal is not a

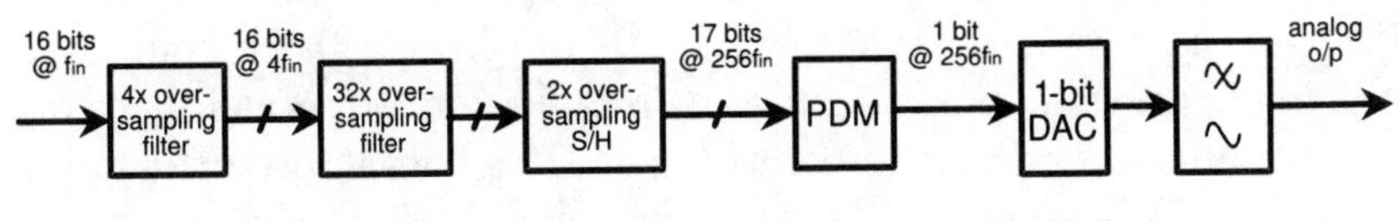


FIGURE 3.49. A commercial bit stream DAC (Philips)

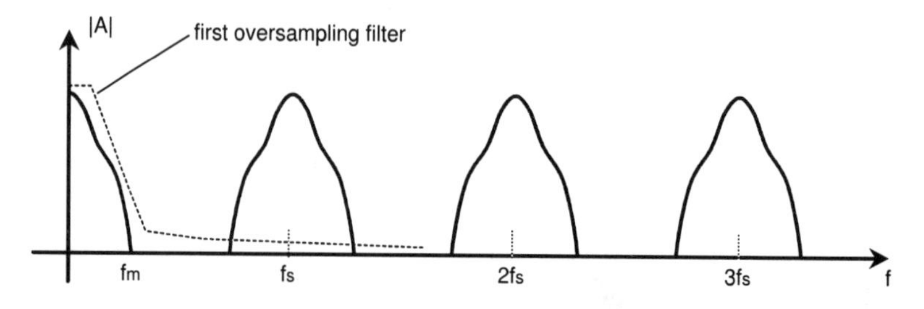


**FIGURE 3.50.** Sampled input waveform and effect of oversampling filter

straight line, so each interpolation stage is followed by a low-pass filter that smooths out the signal, making it closer to the original analog value.

The disadvantage with the system shown in Figure 3.47 is that it requires an 18-bit DAC. We have already seen that this requires extremely good accuracy in the voltage/current source used. This is why in general we see interpolation mixed with bit stream as in the high-performance stereo bit stream DAC from Philips shown in Figure 3.49 (see references at the end of the chapter).

Just one note about this circuit: In this chapter we saw that the spectrum of the sampled input waveform has images of the original signal at multiples of the sampling frequency (Figure 3.50). This is why the first oversampling block in Figure 3.49 is implemented as a filter so that it can also be used to attenuate these images.

## COMMERCIALLY AVAILABLE ADC/DAC ICS

We have covered the different conversion architectures of ADCs and DACs that are available today. We can also look at the different ways these converters connect to DSP devices. There are two basic classes: serial and parallel. A parallel converter receives or produces all the $n$-bits in one pass, as shown in Figure 3.51. Conversely, the serial converters receive or produce their $n$-bits in a serial data stream, as shown in Figure 3.52.

Converters with parallel input and output (I/O) must be attached to the DSP as parallel I/O devices; serial converters can be connected directly to the serial port. Connecting parallel converters as I/O means that we must decode a number of the address lines in order to access both the ADC and the DAC. Also, they must both be attached to the

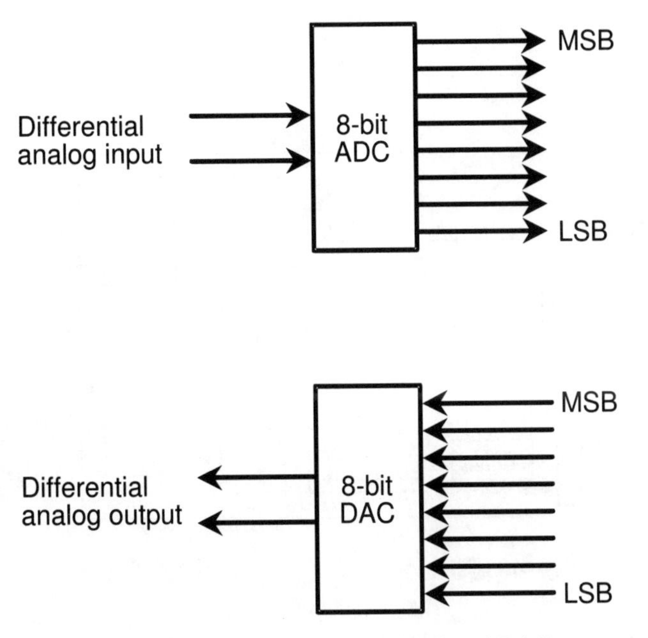


**FIGURE 3.51.** Parallel ADC and DAC

processor's data bus. Although this is not a major stumbling block, we should remember that in most cases we shall also have memory attached to our DSP on these buses. In a typical DSP system we often have many different types of memory (see Chapter 7), slow and fast RAM, EPROM and possibly EEPROM. With all these devices hanging on our data bus, driving the bus may become a problem. This is why many applications are moving toward monolithic serial ADC/

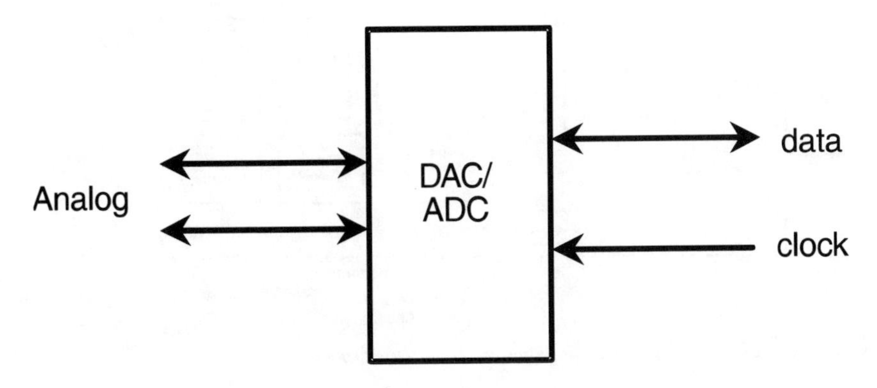


**FIGURE 3.52.** Serial ADC or DAC

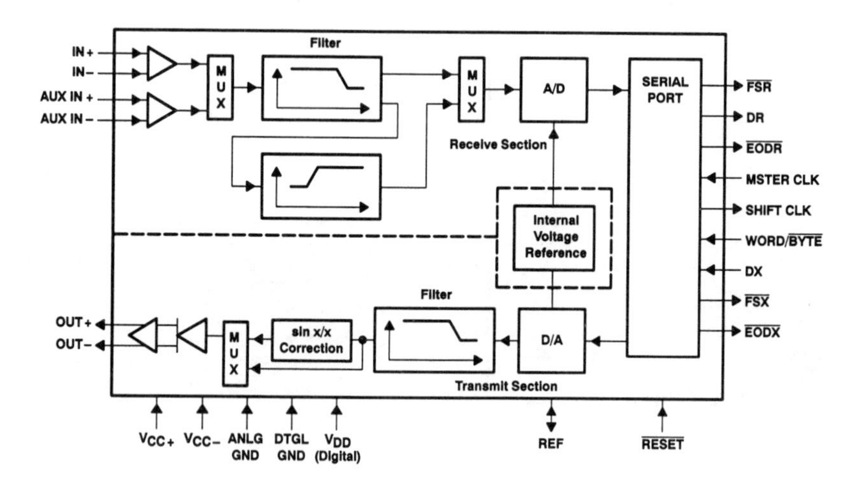


**FIGURE 3.53.** Functional block diagram of TLC32044 analog interface circuit

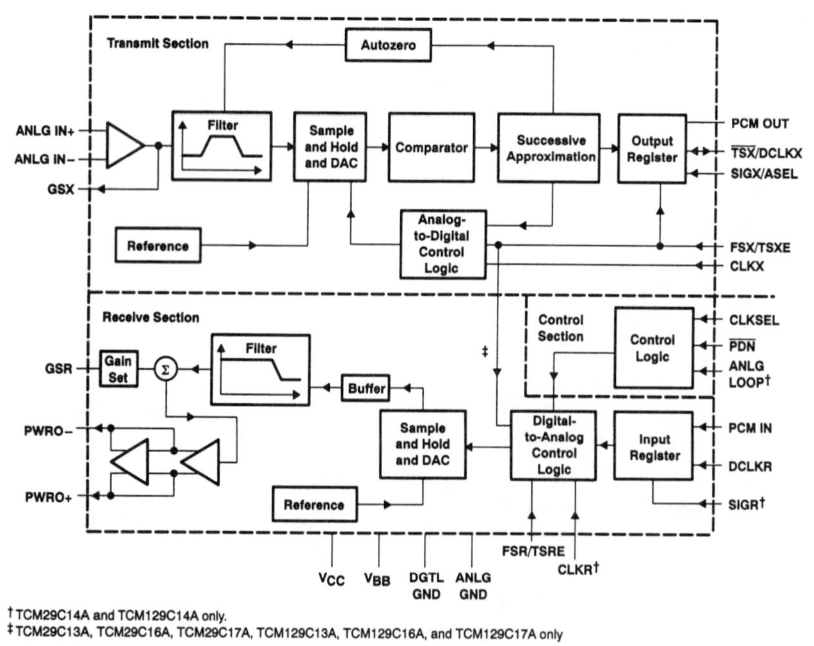


**FIGURE 3.54.** Functional block diagram of typical CODEC device (TCM129C1x)

DACs, i.e., serial devices with both converters on a single piece of silicon. These devices are split into two main types: CODEC and AIC.

Analog Interface Chips (AICs), like the Texas Instruments TLC3204x family, contain all the analog components previously described (Figure 3.53). That is, they contain the ADC and DAC, plus the anti-aliasing and reconstruction filters. The TLC3204x family of AIC filters has programmable cut-off frequencies. Typical applications include modems, speech systems and industrial controllers with sampling rates of between 7.2 and 19.2kHz.

COder/DECoder (CODEC) chips also contain ADC, DAC and filters but their quantizers are usually logarithmic, i.e., A-law or $\mu$-law, depending on whether they are for use in the United States or Europe. Also, the filters in CODECs have a fixed bandwidth, generally 4kHz, revealing that their use is mainly limited to telephony applications (see Figure 3.54).

In addition, as the output of the CODEC is in a nonlinear format, the DSP must convert this into a format it can understand, usually two's-complement (see Chapter 7). Some DSPs implement this function in hardware but if this is not available the DSP can either use a look-up table, or it can calculate the real value at each sample. Similarly, the linear output of the DSP must be reconverted before it is passed back out to the CODEC.

Both CODECs and AICs are designed to interface to the serial port of most general-purpose DSPs. In addition, the AICs can interface with the serial clocks that the DSP produces directly. In the case of CODECs, it is usually necessary to use additional circuitry in order to achieve standard data rates (2400bps, 4800bps, etc).

## DSP SYSTEM ARCHITECTURE

We have now covered all the peripheral circuits required to implement a real-world DSP system. Obviously, in some systems not all of these circuits will be required. For example, DSPs are becoming used as accelerators in mathematical processors, which means that they receive data in a digital format and pass on results in a digital format. In this case we can see that we no longer need an ADC or DAC. Additionally, in control applications the requirements can also be quite different than those for the speech and audio world, but we shall leave such considerations for another time and another book. In

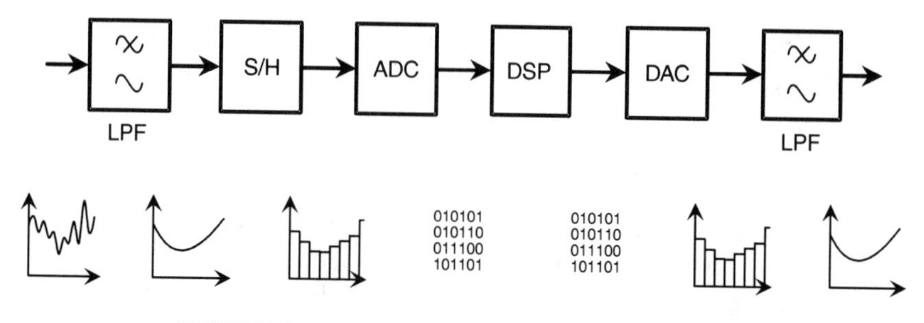


**FIGURE 3.55.** The complete signal processing system

the following chapters we will take this DSP system (Figure 3.55) and look at some of the functions we may wish to perform with it.

## REFERENCES

There are a wide variety of textbooks that detail the conversion between analog and digital signals. The most basic electronic theory books, for example, Zaks [1981] and Horowitz and Hill [1984], give a very clear picture of the operation of most of the standard ADCs and DACs.

The newer forms of converters, i.e., sigma delta ADCs and bit stream DACs, are not nearly so well documented, and the most useful information is contained in papers or manufacturers' data sheets and application notes. Several of these are listed below. A definitive collection of papers can also be found in Candy and Temes (eds.) [1992].

Sampling theory is covered in virtually all textbooks that deal with signal processing, but you may wish to look at the original work of Nyquist [1928] and Shannon [1949]. The paper by Jerri [1977] provides an excellent and comprehensive review of the development of sampling theory and provides many opportunities for further reading. For an up-to-date treatment, refer to Oppenheim and Schafer [1975 and 1988] or to easier text such as Bateman and Yates [1988] and Lynn and Fuerst [1990].

Bateman, A. and Yates, W. [1988]. *Digital Signal Processing Design*, Pitman Publishing, London, UK.

Candy, James C. and Temes, Gabor C. (eds.) [1992]. *Oversampling Delta-Sigma Data Converters*, IEEE Press, New York.

Chu, S. and Burrus, C.S. [Nov 1984]. "Multirate Filter Design Using Comb Filters," *IEEE Transactions on Circuits and Systems,* Vol CAS-31, No 11, pp. 913–924.

Curtis, S. [March 1991]. "Bitstream Conversion," *Electronics World and Wireless World,* Vol 97, No 1661, pp. 205–208.

Finck, R. [June 1989]. "High Performance Stereo Bit-Stream DAC With Digital Filter," *IEEE Transactions on Consumer Electronics,* Vol 35, No 4, pp. 793–796.

Finck, R. and Schulze, W. [May 1990]. "Single Chip CD Decoder," *IEEE Transactions on Consumer Electronics,* Vol 36, No 2, pp. 89–91.

Horowitz, P. and Hill, W. [1984]. *The Art of Electronics,* Cambridge University Press, Cambridge, UK.

Inose, H.; Yasuda, Y. and Marakami, J. [Sept 1962]. "A Telemetering System by Code Modulation, Delta-Sigma Modulation," *IRE Transactions on Space, Electronics and Telemetry,* Set-8, pp. 204–209.

Jerri, Abdul J. [1977]. "The Shannon Sampling Theorem – Its Various Extensions and Applications: A Tutorial Review," *Proceedings of the IEEE,* Vol 65, No 11, pp. 1565–1593, November 1977.

Koch, R. et al. [Dec 1986]. "A 12-bit Sigma-Delta Analog-to-Digital Converter with a 15 MHz Clock Rate," *IEEE Journal of Solid State Circuits,* Vol SC-21, No 6, pp. 1003–1010.

Lynn, P.A. and Fuerst, W. [1990]. *Introductory Digital Signal Processing,* John Wiley and Sons Ltd., Chichester, UK.

Matsuya, Y. et al. [Dec 1987]. "A 16-bit Oversampling A-to-D Conversion Technology Using Triple-Integration Noise Shaping," *IEEE Journal of Solid State Circuits,* Vol SC-22, No 6, pp. 921–929.

Naus, P.J.A. et al. [June 1987]. "A CMOS 16-bit D/A Converter for Digital Audio," *IEEE Journal on Solid State Circuits,* SC-22, No 3.

Nyquist, H. [1928]. "Certain Topics in Telegraph Transmission Theory," *AIEE Transactions,* pp. 617–644.

Oppenheim, A.V. and Schafer, R.W. [1975 and 1988]. *Digital Signal Processing,* Prentice Hall, Englewood Cliffs, NJ.

Park, S. [1990]. *Principles of Sigma-Delta Modulation for Analog-to-Digital Converters,* Motorola Inc.

Rebeschini, M. et al. [1989]. "A High-Resolution CMOS Sigma-Delta A/D Converter With 320kHz Output Rate," *Proceedings ISCAS,* pp. 246–249.

Shannon, C. E. [1949]. "Communications in the Presence of Noise," *Proceedings of the IRE,* Vol 37, pp. 10–21, January 1949.

Steele, R. [1975]. *Delta Modulation Systems,* Pentech Press, London.

Stewart, R.W. [1991]. "Digital Signal Processing: Technology and Marketing for Audio Systems," *IEE Colloquium on Digital Audio Signal Processing,* Digest No 107, pp. 5/1–6.

Thompson, C.D. [May 1989]. "A VLSI Sigma-Delta A/D Converter For Audio and Signal Processing Applications" *Proceedings of ICASSP,* Glasgow, UK.

Welland, D.R. et al. [Nov 1988]. "A Stereo 16-bit Sigma-Delta A/D Converter For Digital Audio," *Proceedings of the 85th Convention of the Audio Engineering Society,* Vol 2724, H-12, California.

Zaks, R. [1981]. *From Chips to Systems,* Sybex, California.

# 4

# Filtering

In this chapter we look at filtering, first through a brief review of analog filters, followed by a description of how we can perform the same functions in the digital domain. We shall also cover some of the software tools available that help us to design digital filters.

## WHAT IS FILTERING?

We know from basic electronic theory that capacitors and inductors have impedances ($X$) that depend on frequency, as defined below:

$$X_c = \frac{1}{j\omega C} \qquad C = \text{capacitance (Farads)}$$

$$X_L = j\omega L \qquad L = \text{inductance (Henrys)}$$

where

$$\omega = 2\pi f \qquad f = \text{frequency (Hertz)}$$

and

$$j = \sqrt{-1}$$

When either component is combined with a resistor we can build frequency-dependent voltage dividers. For example, let's look at a

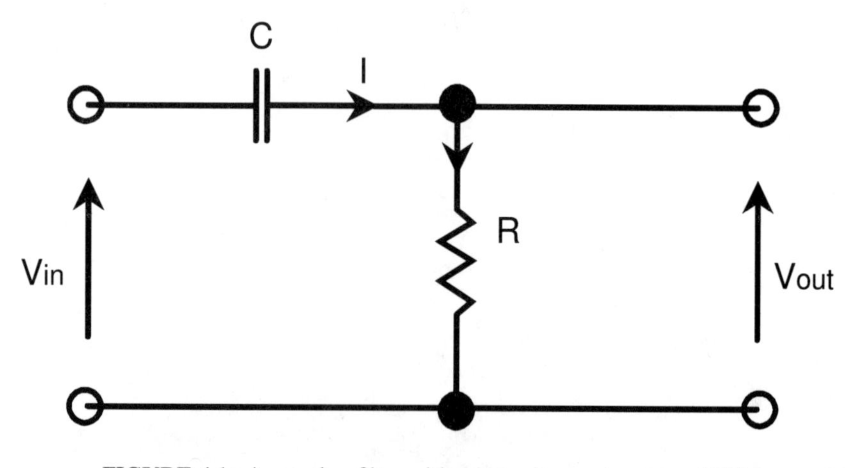


**FIGURE 4.1.** An analog filter with a capacitor and resistor (CR)

circuit containing a resistor and a capacitor as shown in Figure 4.1. If we now apply Ohm's law to this circuit we can see that:

$$V_{in} = I(R + X_c)$$

so that:

$$I = \frac{V_{in}}{R + \frac{1}{j\omega C}}$$

Also

$$V_{out} = IR$$

so solving for $I$ gives:

$$H(\omega) = \frac{V_{out}}{V_{in}} = \frac{R}{R + \frac{1}{j\omega C}}$$

The resulting transfer function ($H(\omega)$) has both phase and gain components. These can be found using the following equations (see Figure 4.2):

$$\text{Gain} = |\,A\,| = \sqrt{\Re H(\omega)^2 + \Im H(\omega)^2}$$

$$\text{Phase} = \ \phi \ = \tan^{-1}\frac{\Im H(\omega)}{\Re H(\omega)}$$

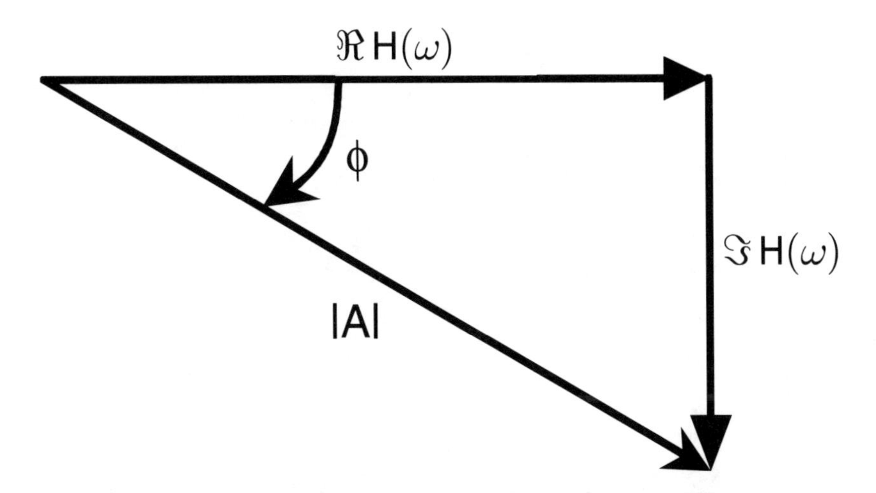


**FIGURE 4.2.** Phase and amplitude components of the $H(\omega)$ transfer function

For those of you unfamiliar with these terms, a complex function is defined in the following way:

$$H(\omega) = \Re H(\omega) + j\Im H(\omega)$$

where

$$\Re H(\omega) = \text{Real part of } H(\omega)$$

and

$$\Im H(\omega) = \text{Imaginary part of } H(\omega)$$

From this we can work out that the gain of our voltage divider is given by:

$$|A| = \frac{R}{\sqrt{R^2 + \dfrac{1}{(\omega C)^2}}}$$

If we plot this against frequency we obtain a response as shown in Figure 4.3. We can see that for a constant input voltage, at high frequencies the output is approximately equal to the input and at low frequencies the output tends towards, zero. This circuit is called a high-pass filter because it only allows high frequencies to pass unattenuated.

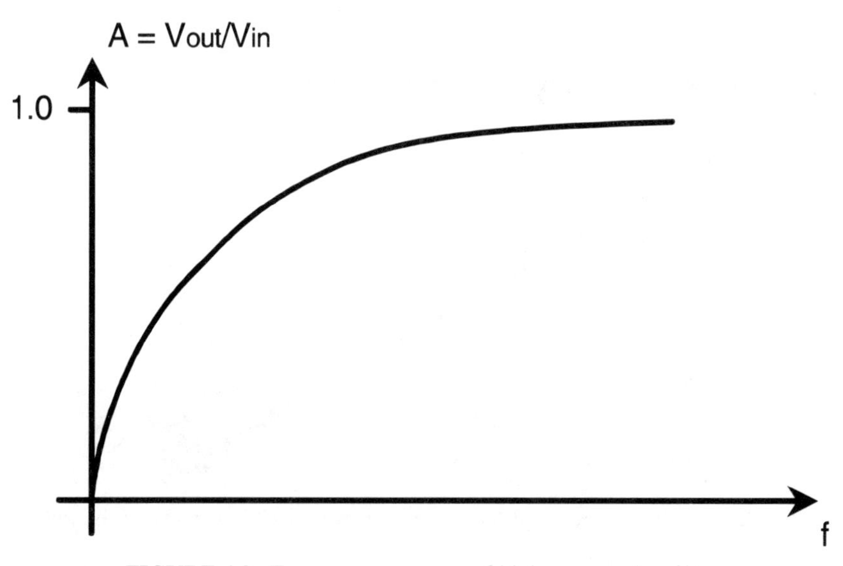


**FIGURE 4.3.** Frequency response of high-pass analog filter

If we now turn the circuit around as shown in Figure 4.4 and again derive the gain response of the circuit we find:

$$|A| = \frac{\frac{1}{|\omega|C}}{\sqrt{R^2 + \frac{1}{(\omega C)^2}}} = \frac{1}{\sqrt{R^2C^2\omega^2 + 1}}$$

This gain response with frequency is shown in Figure 4.5. We can see that at low frequencies the output is approximately equal to the input and at high frequencies the output tends toward zero. Hence, we have a low-pass circuit, i.e., it only allows low frequencies to pass unattenuated.

This mathematical treatment can be repeated with inductors but we leave that for you to try yourselves if you want to. In general, capacitors and resistors are used to design analog filters since inductors are bulky in comparison, more expensive and do not perform as well. Real inductors also contain parasitic resistance and capacitance causing their actual amplitude response to depart significantly from the ideal.

The exception to this is the use of ferrite beads in high-frequency circuits. A ferrite bead on a wire makes it slightly inductive, producing

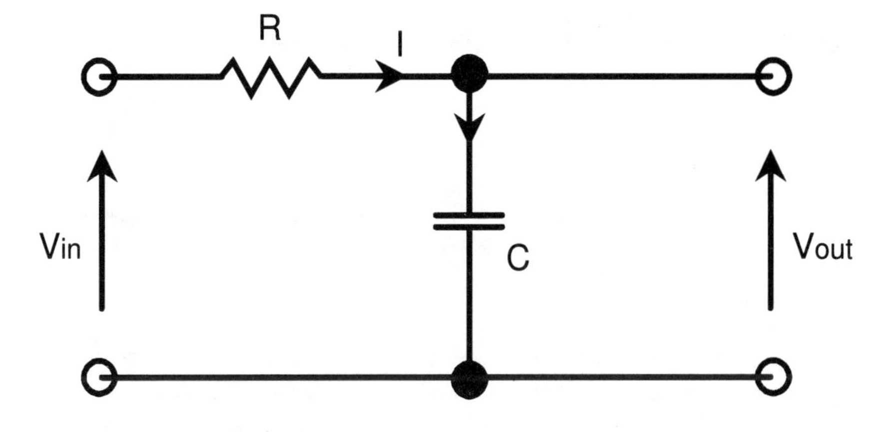


**FIGURE 4.4.** An analog filter with a resistor and capacitor (RC)

an appreciable impedance at high frequencies. This impedance is used to prevent unwanted oscillations occurring at high frequencies.

Simple RC filter structures, as described above, produce filters with a gentle roll-off in amplitude of around 6dB per octave. To achieve faster rates of roll-off it is possible to cascade such filter sections into a structure called a *ladder filter* (Figure 4.6). For each additional RC circuit we increase the roll-off by 6dB per octave. For example, a three-stage network will have an 18dB roll-off. It is interesting to note that one method of describing filters is to refer to the number of RC sections they contain as the "order" of the filter: A three-stage filter would be third order.

From this, we might mistakenly believe that designing an *n*th-order

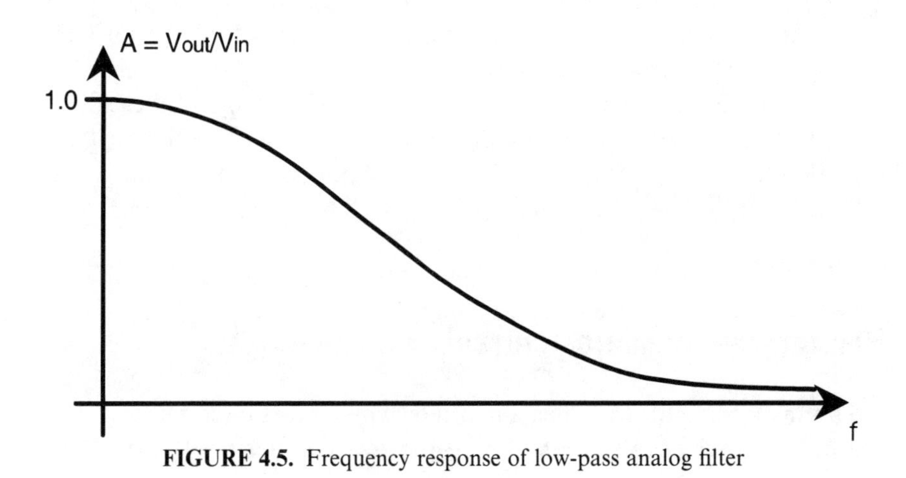


**FIGURE 4.5.** Frequency response of low-pass analog filter

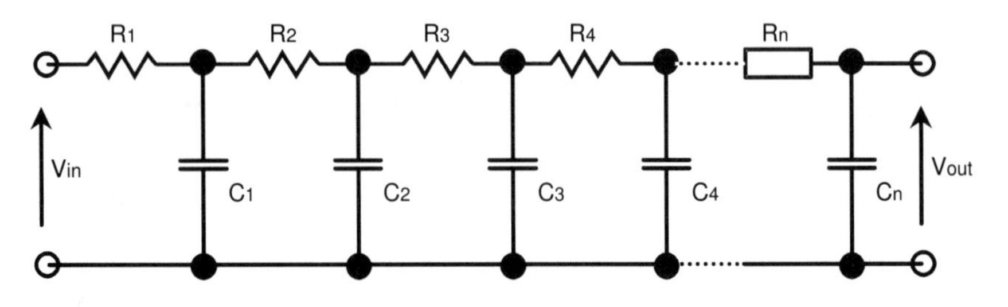


**FIGURE 4.6.** A ladder filterstructure

filter is as simple as cascading *n* similar RC sections. Unfortunately, this isn't the case. The original analysis of a first-order circuit assumes that the impedance of the driving voltage source is zero and the impedance of the load is infinite. Consequently, once we start adding more stages we alter the source and load impedances of each stage significantly, and we must therefore recalculate the frequency response of the whole design.

Nowadays, we use computer programs to calculate the correct resistor and capacitor values for our desired gain and phase response. Even then it is customary to assume our components are ideal, i.e., they have no parasitic elements. This is why we see variable resistors and capacitors in analog filters. It is usually necessary to "tweak" the circuit in order to produce the final desired result. In addition, the characteristics of all discrete components drift with temperature and time, meaning that it is sometimes necessary to retune the circuit even during its use.

An extension to this discrete filter circuit is the use of operational amplifiers to produce what are known as "active" filters. The main advantage of the operational amplifier approach is that it allows the design of filters with much faster roll-off rates. Although this is an interesting topic in itself we shall not examine the design of these filters in this book; extensive information can be found in virtually any basic textbook on electronics, for example, Zaks [1981] or Ahmed and Spreadbury [1978].

## FILTER PERFORMANCE CRITERIA

This leads us to how we measure filters. There must be some agreed-on criteria by which we can compare the relative merits of two similar

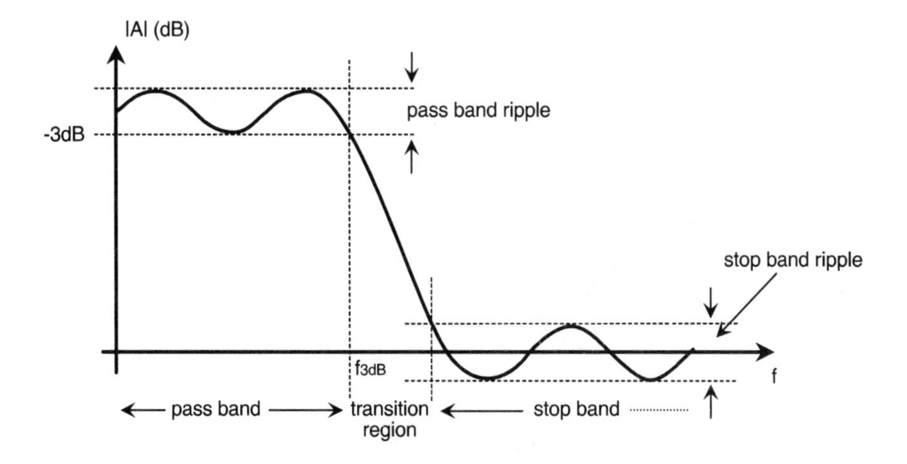


**FIGURE 4.7.** The amplitude response and performance measurements of a filter

filters. In general the gain response of the filter, as shown in Figure 4.7, is used. The figure indicates a typical characteristic for a low-pass filter.

The pass band is defined as the range of frequencies over which signals pass virtually unattenuated through the circuit. The pass band extends to the point where the response drops off by 3dB, which is known as the cut-off frequency ($f_{3dB}$). It is possible to design filters that have no ripple over this pass band (see later), but we usually accept a certain level of ripple in this region in exchange for a faster roll-off of gain with frequency in the transition region.

The transition region is the area between the pass band and stop band. We have already referred to the roll-off in this band. This rate of change of gain with frequency is another important filter performance criterion.

The stop band is chosen by the designer depending on his requirements. For example, it may be defined as where the gain response falls below −40dB. In another application, the stop band may be where the filter response falls below −80dB. The important feature is that the response throughout the stop band should always be below the design specification. There are several types of filter (e.g. the equiripple) that have ripple in the stop band. As long as this ripple is below the required stop band level it is of little importance.

It is unrealistic to describe a filter solely by the variation of its gain with frequency. The other important characteristic is its phase response (Figure 4.8). If we remember an earlier equation we saw

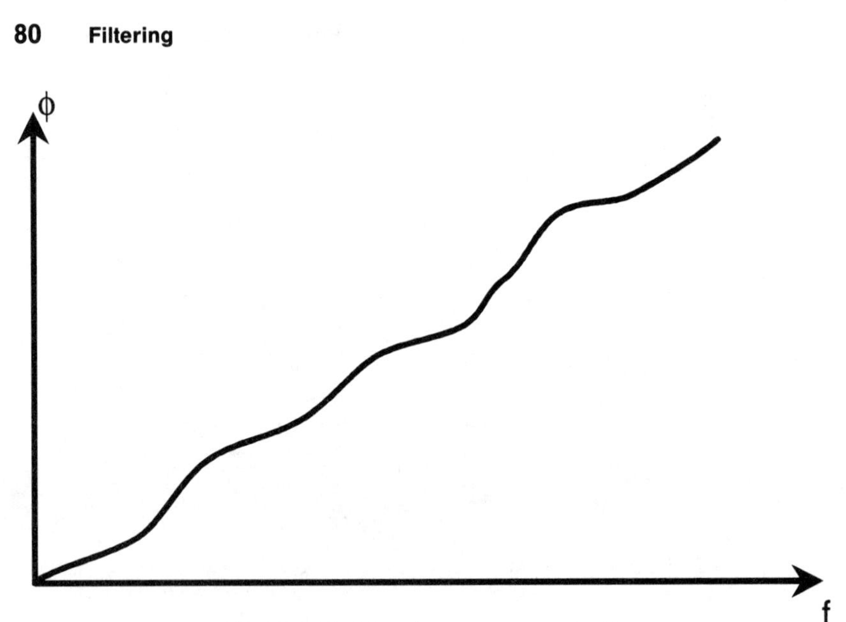


**FIGURE 4.8.** The phase response of a filter

that phase response was given by:

$$\phi = \tan^{-1} \frac{\Im H(\omega)}{\Re H(\omega)}$$

Phase is important because it is directly related to the time delay of the different frequencies passing through the filter. For example, a filter

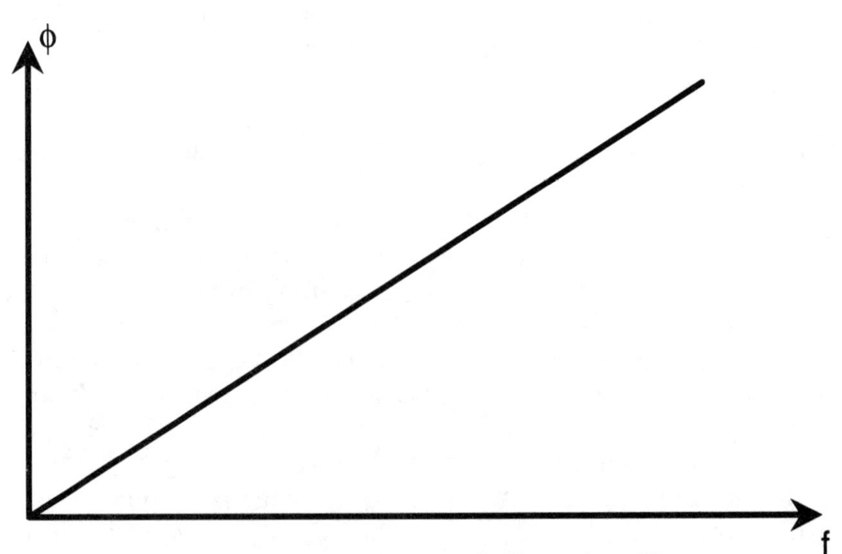


**FIGURE 4.9.** Phase response of a linear phase filter

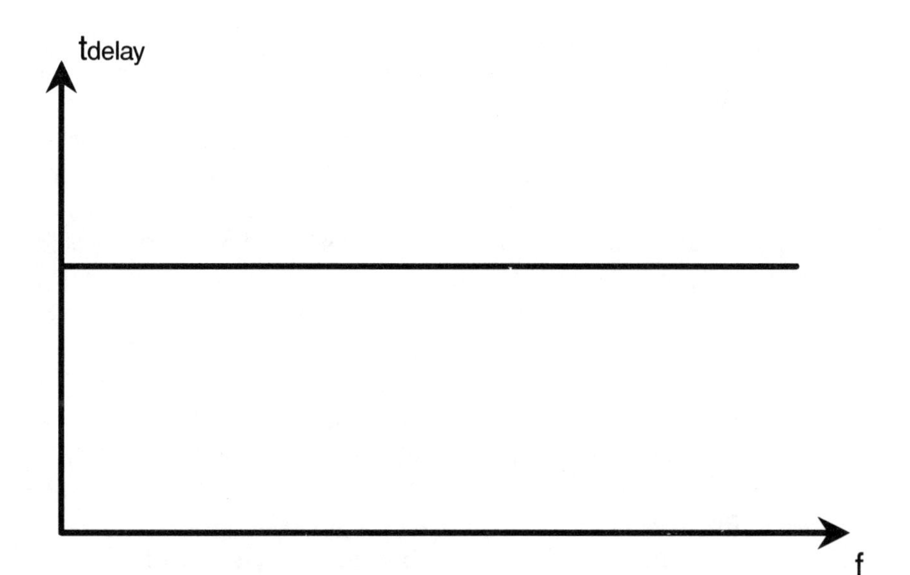


**FIGURE 4.10.** Uniform time delay of a linear phase filter

with a linear phase response (Figure 4.9) delays all frequencies by the same amount of time (Figure 4.10). Conversely, a filter with a nonlinear phase response causes all frequencies to be delayed by different periods of time, meaning that the signal emerges distorted at the filter output. In

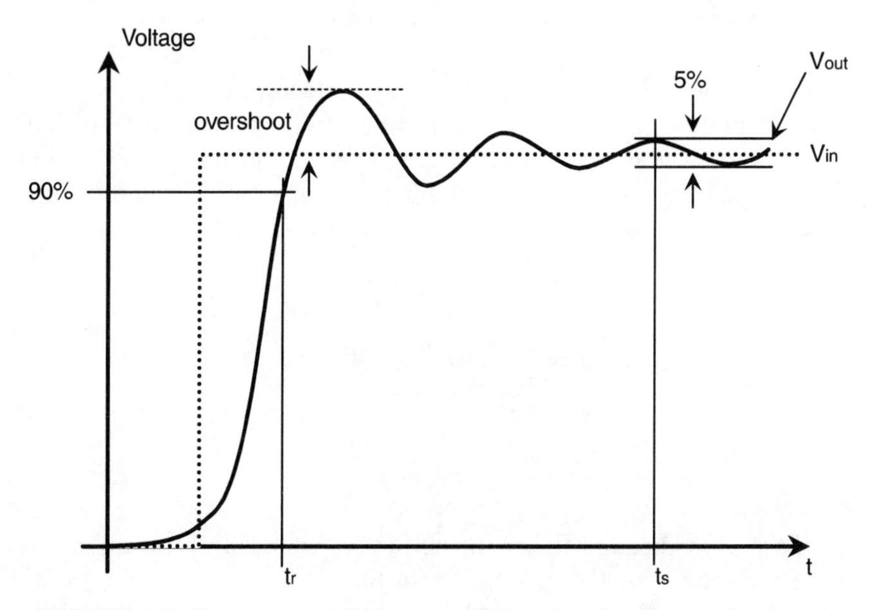


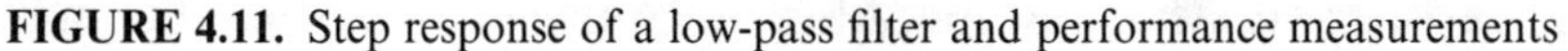


**FIGURE 4.11.** Step response of a low-pass filter and performance measurements

radio-frequency (RF) applications we are familiar with the "scattering" of a radio signal due to the different and varying path lengths. This effect is very similar to our example of nonlinear phase in a filter.

If we are to be realistic when designing filters, linear phase is actually important only over the pass band and transition band of our filter since all frequencies beyond that are attenuated. Also, as most designs are a compromise, we often live with small variations in the phase response in order to obtain better performance in some other characteristic, e.g., the filter roll-off or filter order.

We can also describe a filter using its time domain response. Figure 4.11 shows a typical response for a low-pass filter when a sudden step change in voltage is applied to its input.

The common measures of performance for the response of the filter in the time domain are:

Rise time: The time taken for the output to reach 90% of full scale ($t_r$).

Settling time: The time taken to settle within 5% of the final value ($t_s$).

Overshoot: The maximum amount by which the output momentarily exceeds the desired value after level transition.

Ringing: Oscillations about the final (mean) value.

Let's now examine the response of a simple low pass filter over time when a number of step changes in input are applied to it. We shall repeat the same example later, but with a digital filter. The analog filter we shall use is shown in Figure 4.12, our original crude voltage divider. The time constant, $\tau$, of the circuit is given by:

$$\tau = CR$$

$$\tau = 10 \times 10^{-9} \times 100 \times 10^{3}$$

$$\tau = 10^{-3}$$

$$\tau = 1\text{ms}$$

If we plod through the mathematics we should find that the time constant represents the time taken for the capacitor to reach approximately 63% of its final value. Also, we could derive the fact that in five

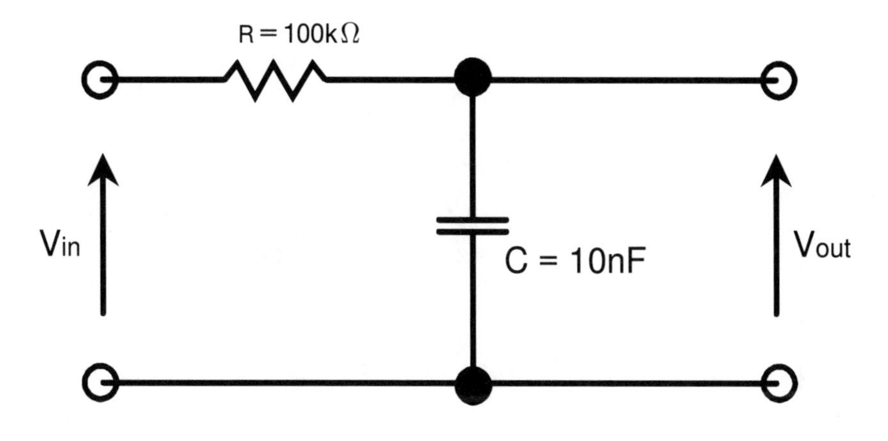


**FIGURE 4.12.** Simple RC analog low-pass filter

times the time constant the capacitor will be charged to within 1% of its final voltage. We can use these two figures to calculate the output of the filter.

Let's now look at what happens when we apply the input signal shown in Figure 4.13 to the circuit. If we assume that the capacitor is initially discharged, during the first period the capacitor voltage will

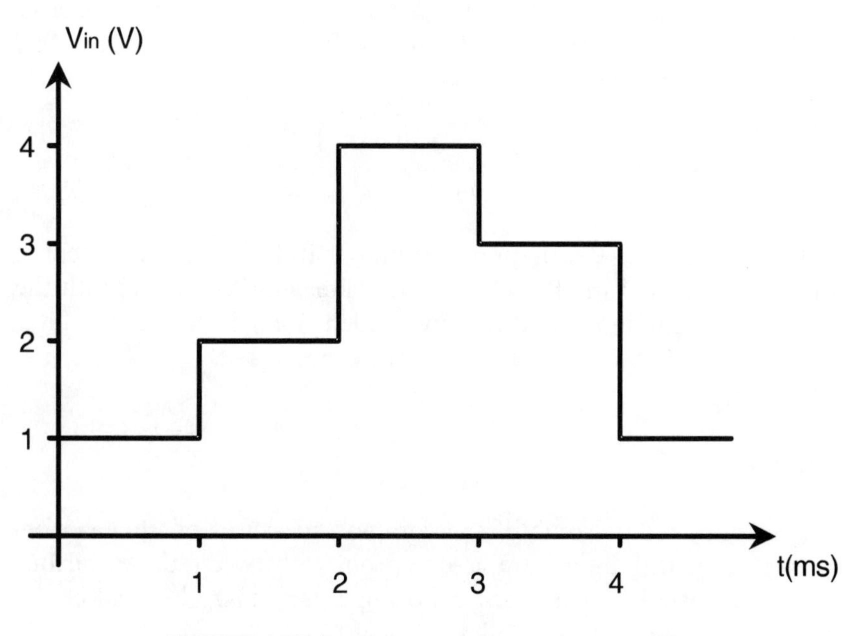


**FIGURE 4.13.** Step input waveform to low-pass filter

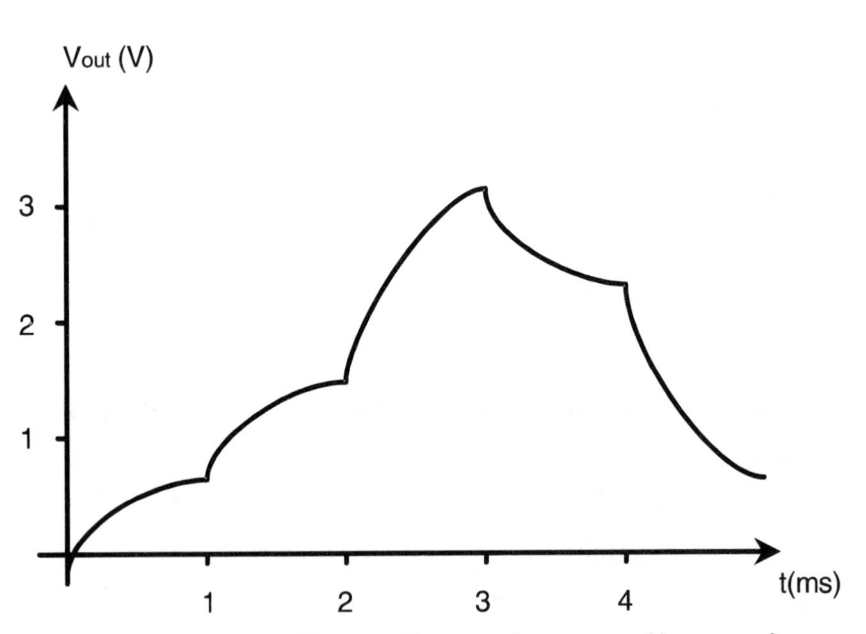


**FIGURE 4.14.** Output of low-pass filter - moving average of input waveform

ramp up exponentially toward 1V, and after a period of 1ms, $V_{out}$ will have reached 0.63V. At this time the input signal jumps to 2V so that the difference between desired and actual $V_{out}$ is 1.37V. The capacitor begins its charging again with this new voltage and at the end of 2ms $V_{out}$ will have reached:

$$V_{out} = 0.63 + (0.63 \times 1.37)$$
$$= 1.49\text{V}$$

Repeating these calculations we find that $V_{out}$ follows the curve shown in Figure 4.14. The effect of the filter has been to smooth the input waveform and round off any sudden changes. In fact the filter has performed a moving average of the input signal.

## FILTER TYPES

We have looked at some of the common measures of filter performance. Practical filters are a compromise between these various measurements. The reason for choosing a particular filter will be the relative importance of the different measurements.

Let's go back to the gain response of a filter shown in Figure 4.7. We saw that the main characteristics are pass-band ripple, cut-off frequency ($f_{3dB}$), roll-off and stop-band attenuation. To this we can also add the phase response. There are a number of filter types that optimize one or more of these characteristics and are common to both analog and digital designs. You will find these filter styles referred to many times in textbooks, and computer-based filter design packages usually offer a choice of filter types to create. We shall describe only the four most commonly used filter types and the pros and cons of each. For a detailed description of the many types of filter, refer to Jackson [1989].

### Butterworth Filter

The Butterworth filter has a completely flat gain response over the pass band and it is often referred to as the "maximally flat" filter. This flat pass band is achieved at the expense of the transition region, which has a very slow roll-off, and the phase response, which is nonlinear around the cut-off frequency. The gain of this type of filter is given by:

$$\left|\frac{V_{out}}{V_{in}}\right| = \frac{1}{\sqrt{1+\left(\frac{f}{f_{3dB}}\right)^{2n}}}$$

where $n$ is the order of the filter. If we increase the order of the filter the flat region of the pass band gets closer to the cut-off frequency before it drops away and we can improve the roll-off (Figure 4.15). Referring back to the original ladder filter, remember that increasing the order of the filter also increases the complexity of its design.

The Butterworth filter may appear to be an ideal solution for all our designs, but more often than not it falls well below our circuit requirements, either because we must use an excessively high-order filter to achieve our specified roll-off, or because of its relatively poor phase response. In most applications we can tolerate a controlled amount of ripple in the pass band in exchange for a much steeper roll-off in the transition region.

### Chebyshev Filter

The Chebyshev filter provides this exchange: It permits a certain amount of ripple in the pass band but has a very much steeper

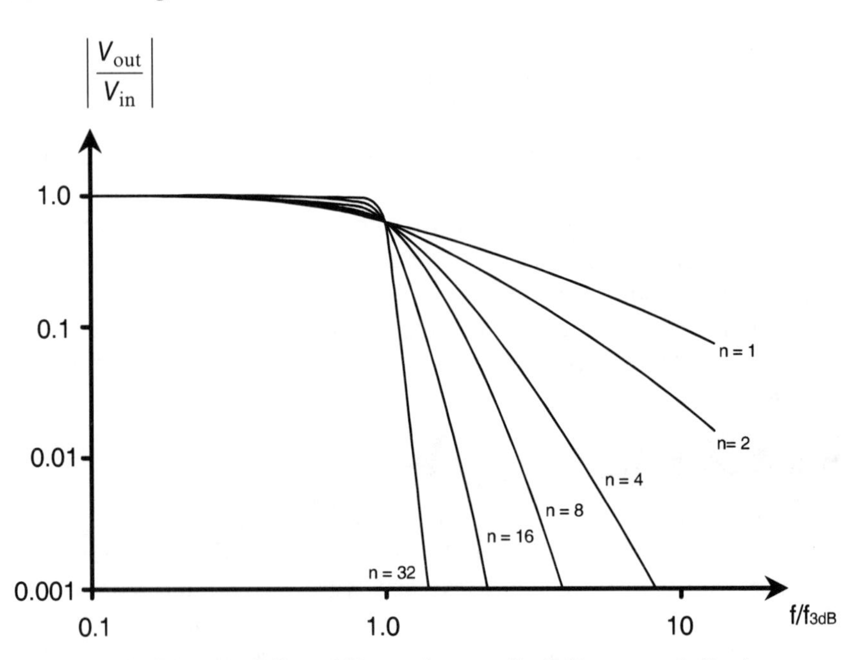


**FIGURE 4.15.** Effect of filter order on roll-off (Butterworth filter)

roll-off. The gain response of a Chebyshev filter is given by:

$$\left|\frac{V_{out}}{V_{in}}\right| = \frac{1}{\sqrt{1 + \epsilon^2 C_n^2\left(\frac{f}{f_{3dB}}\right)}}$$

where $C_n$ is a special polynomial that is a function of $n$ (the order of the filter) and $\epsilon$ is a constant that determines the amount of ripple in the pass band. The gain response of a Chebyshev filter is shown in Figure 4.16. Even if we place very tight limits on the pass-band ripple, the improvement in roll-off is considerable when compared with the Butterworth filter. It can be shown that for a pass-band flatness within 0.1dB and a stop-band attenuation of 20dB 25% beyond the cut-off frequency ($f_{3dB}$), we need an eighth-order Chebyshev filter or a nineteenth-order Butterworth filter.

Although the Chebyshev filter is an improvement on the Butterworth filter with respect to the roll-off, they both have poor phase responses, the Chebyshev's being worse. The Chebyshev filter style is sometimes called the equiripple filter, as the ripples are always of equal size throughout the pass band. In addition, the number of ripples increases with the order of the filter.

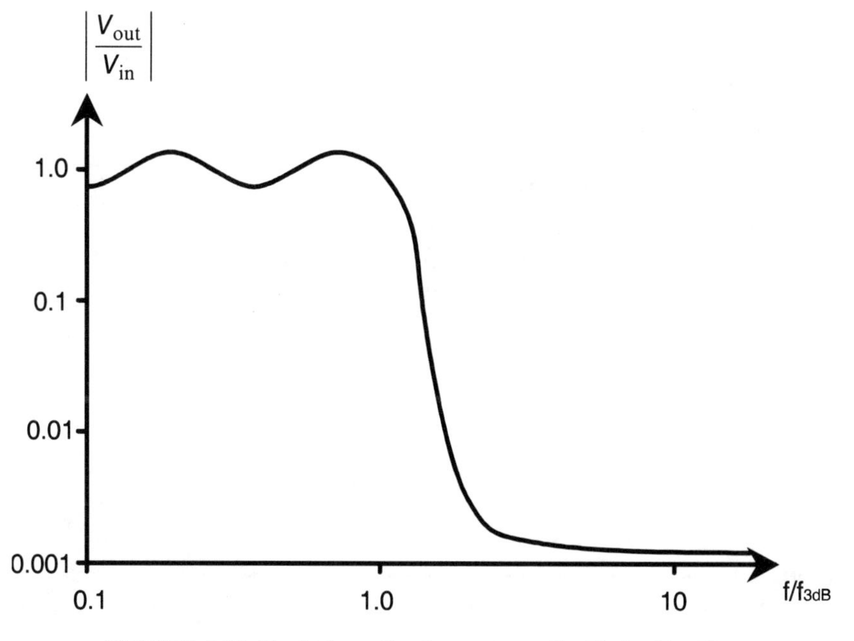


**FIGURE 4.16.** Typical amplitude response of a Chebyshev filter

### Elliptic Filter

The elliptic, or Cauer, filter is an extension of the trade-off between ripple in the pass band and roll-off. The gain response of the elliptic filter is given by the following deceptively simple expression:

$$H(\omega) = \frac{1}{\sqrt{1 + \epsilon^2 U_n^2\left(\frac{f}{f_{3dB}}\right)}}$$

The function $U_n(f/f_{3dB})$ is called the Jacobian elliptic function. The discussion of this filter, even on a superficial level, is far beyond the scope of this book. Suffice it to say that elliptic filters achieve the maximum possible roll-off for a particular filter order. The phase response of an elliptic filter is extremely nonlinear, so we can only use this design where phase is not an important design parameter. References for further reading on elliptic filters can be found at the end of this chapter.

### Bessel filter

By now it will be clear that the gain response is only part of the story

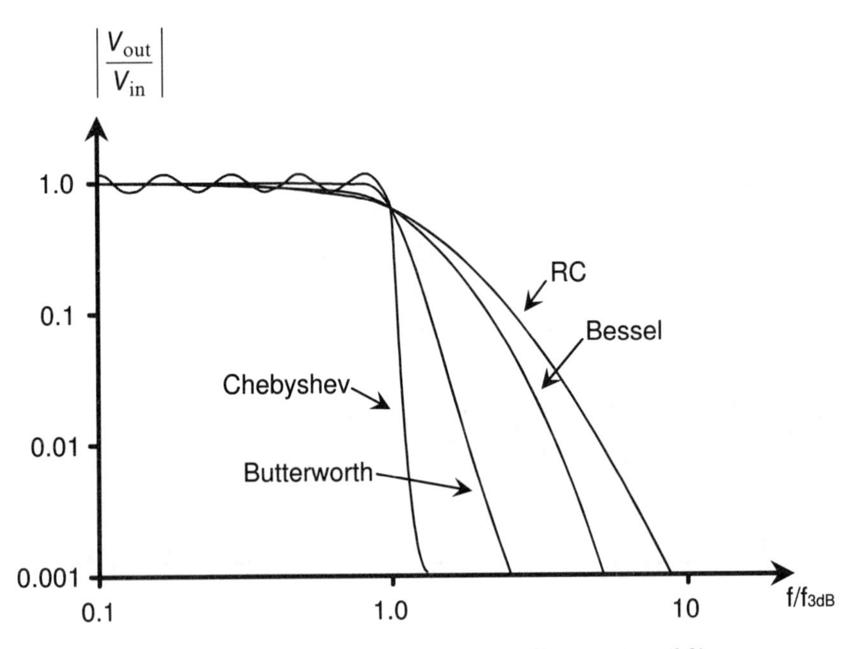


**FIGURE 4.17.** Amplitude response of different types of filter

where filters are concerned. A poor phase response is undesirable in many applications, the most obvious example being audio systems. A badly designed filter in such cases could be disastrous. Enter the Bessel, or Thomson, filter, which has a maximally flat phase response in its pass band.

The gain response of the Bessel function is compared with the same-order RC, Butterworth and Chebyshev filters in Figure 4.17. Figure 4.18 shows the time domain response of Bessel and Butterworth filters of the same order. From these two figures it can easily be seen that to obtain a good phase response we must sacrifice steepness in the transition region.

There are many other types of filter design that allow a compromise between steepness, ripple and phase response. Here we have seen the extremes – the elliptic with its good gain response and the Bessel with its excellent phase response. Let's now say a kind word about the Butterworth filter and explain why it was included here – it is a good compromise of all the important features in filters.

As we said in the preamble to this short section on the major filter types, the choice of filter rests with the designers. It is for them to decide which one best fits the requirements of the design.

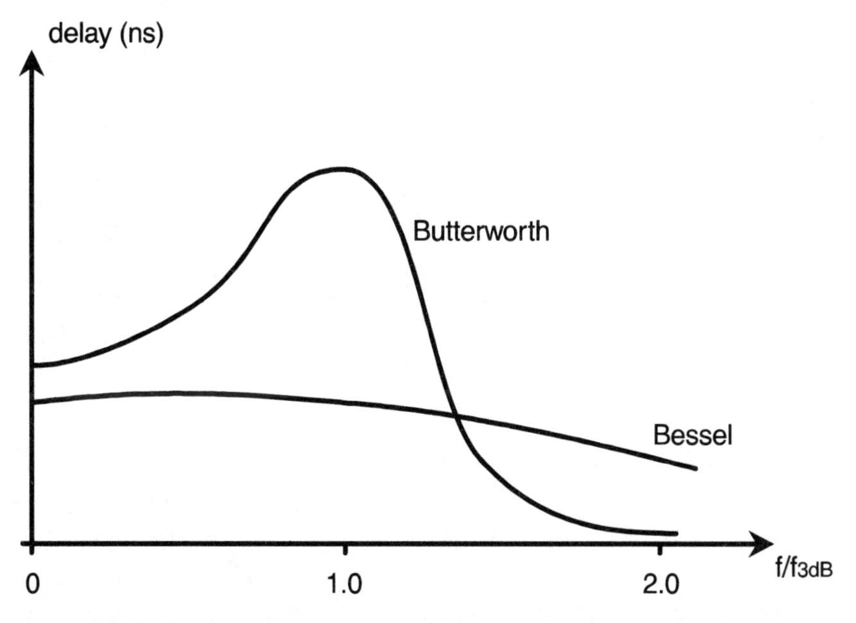


**FIGURE 4.18.** Time delay response of Bessel and Butterworth filters

Computer-based filter design packages have made it far easier to experiment with different filter types, specifications and filter orders. Magnitude, phase and impulse response can be evaluated in a few seconds. We shall say a little more about filter design programs later.

## DIGITAL FILTERS

For many years digital filters have been the most common application of digital signal processors. As we already saw in Chapter 2, digitizing any design ensures that we can reproduce it time and time again with exactly the same characteristics. There are two other significant advantages with respect to filters. First, it is possible to reprogram the DSP and drastically alter the filter's gain or phase response. For example, we can reprogram a system from low pass to high pass without throwing away our existing hardware. Second, we can update the filter coefficients while the program is running, i.e., we can build "adaptive" filters.

If you read more in-depth books on filters, you will find them full of mathematical functions called Z-transforms that they use to describe the workings of digital filters. This book is not intended as a textbook on mathematics so we shall not attempt to explain these transforms. It will be sufficient for our purposes simply to understand the implications of the transforms and look at how we implement filters with present-day DSPs. For the enthusiasts among you, Jury [1964] is the definitive work on the Z-transform and most DSP textbooks cover the subject comprehensively.

There are two basic forms of digital filter: the Finite Impulse Response (FIR) filter and the Infinite Impulse Response (IIR) filter, which are explained below. The initial descriptions are based on a low-pass filter but it is very easy to change low-pass filters to any other type – high-pass, band-pass, etc. Parks and Burrus [1987] and Oppenheim and Schafer [1975 and 1988] cover this in detail.

### Finite Impulse Response Filters

The block diagram of a Finite Impulse Response (FIR) filter is shown in Figure 4.19. The input signal $x(n)$ is a series of discrete values obtained by sampling an analog waveform. In this series $x(0)$ corresponds to the input value at $t = 0$, $x(1)$ is the value at $t = t_s$, $x(2)$ is the value at $t = 2t_s$, and so on. The value $t_s$ is the sampling period, where:

$$t_s = \frac{1}{f_s}$$

The "$z$" in Figure 4.19 is the Z-transform which we referred to earlier. For our purposes we shall simply think of the block shown as $z^{-1}$ as being a time delay of one sampling period ($t_s$), also known as the *unit delay*. We can then see that the value shown as $x(n - 1)$ is actually the

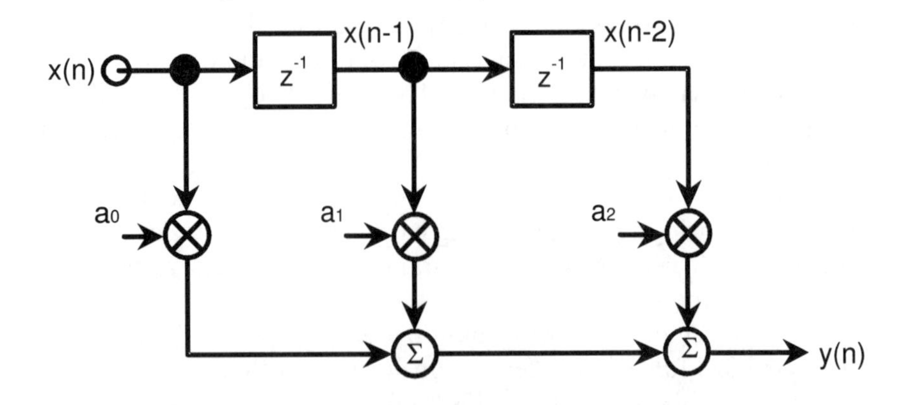


**FIGURE 4.19.** Block diagram of a finite impulse response (FIR) filter

value of $x(n)$ one time period before now, i.e., the previous input. Similarly, $x(n-2)$ is the value of the input two sampling periods beforehand. This simple fact makes the FIR filter circuit extremely easy to understand.

In the example shown in Figure 4.19, the output signal $y(n)$ is therefore always a combination of the last three input samples. In the diagram each of the samples is multiplied by a coefficient, $a_R$, to give:

$$y(n) = a_0x(n) + a_1x(n-1) + a_2x(n-2)$$

In order to demonstrate the operation of this filter, let's look at a "real" example. The following table lists the end-of-day share prices over a period of a week. We shall use these values as our input samples, i.e., $x(n)$. The value for $n = 0$ is taken to be the price of shares on Monday:

| Day | Period | $x(n)$ | Price ($) |
|---|---|---|---|
| Monday | 0 | $x(0)$ | 20 |
| Tuesday | 1 | $x(1)$ | 20 |
| Wednesday | 2 | $x(2)$ | 20 |
| Thursday | 3 | $x(3)$ | 12 |
| Friday | 4 | $x(4)$ | 40 |
| Saturday | 5 | $x(5)$ | 20 |
| Sunday | 6 | $x(6)$ | 20 |

Plotting a histogram of the above values against time gives Figure 4.20.

We shall take the FIR filter shown in Figure 4.19 with the following values for the $a_R$ coefficients:

| $a_R$ | Value |
|---|---|
| $a_0$ | 0.25 |
| $a_1$ | 0.50 |
| $a_2$ | 0.25 |

The filter now looks like Figure 4.21. When $n = 0$ the first value of $x(n)$, i.e., $x(0)$, is applied at the input to the circuit. Assuming that there have been no share prices for the two previous days, the values of

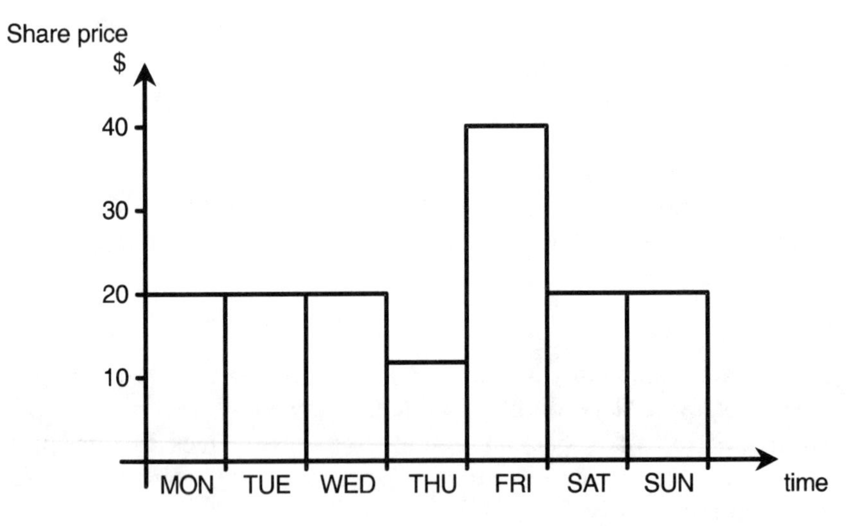


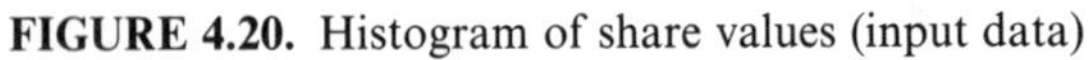
**FIGURE 4.20.** Histogram of share values (input data)

$x(n)$, $x(n-1)$, and $x(n-2)$ at $n = 0$ are:

$$x(0) \quad = 20$$
$$x(-1) = 0$$
$$x(-2) = 0$$

Performing the multiplications and additions, we obtain the value of the output:

$$\begin{aligned} y(0) &= 0.25 \times 20 + 0.5 \times 0 + 0.25 \times 0 \\ &= 5 \end{aligned}$$

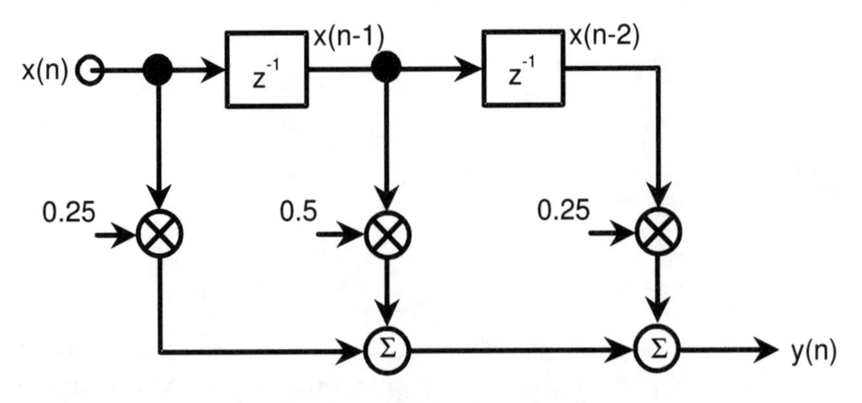


**FIGURE 4.21.** FIR showing values for $a_R$

Moving on to the next sample period, Tuesday's share price is the input value and Monday's is displaced to $x(n-1)$, i.e., it has become the old value, one sampling period behind. In this example the sampling period $t_s$ = one day, so $n = 1$ and the values for $x(n)$, $x(n-1)$ and $x(n-2)$ are:

$$x(1) = 20$$
$$x(0) = 20$$
$$x(-1) = 0$$

It follows that:

$$\begin{aligned} y(1) &= 0.25 \times 20 + 0.5 \times 20 + 0.25 \times 0 \\ &= 15 \end{aligned}$$

Moving on again to Wednesday, $n = 2$ so:

$$x(2) = 20$$
$$x(1) = 20$$
$$x(0) = 20$$

giving:

$$\begin{aligned} y(2) &= 0.25 \times 20 + 0.5 \times 20 + 0.25 \times 20 \\ &= 20 \end{aligned}$$

For Thursday, $n = 3$:

$$x(3) = 12$$
$$x(2) = 20$$
$$x(1) = 20$$

Notice that Monday's share price has dropped out of the equation. We shall come back to this aspect shortly. The output value for Thursday is given by:

$$\begin{aligned} y(3) &= 0.25 \times 12 + 0.5 \times 20 + 0.25 \times 20 \\ &= 18 \end{aligned}$$

Repeating the calculation for all the data up to Sunday ($n = 6$), we obtain the values for $y(n)$ as given below:

| Day | $y(n)$ |
|---|---|
| Monday | 5 |
| Tuesday | 15 |
| Wednesday | 20 |
| Thursday | 18 |
| Friday | 21 |
| Saturday | 28 |
| Sunday | 25 |

We can plot these output values as shown in Figure 4.22. We have assumed that the output from the circuit would be passed through a DAC and then a reconstruction filter (see Chapter 3) and have joined up the points with straight lines. We can see that the filter is performing a moving average calculation, similar to the action of the analog filter discussed earlier in this chapter.

There are a couple of terms that are peculiar to digital filters. The string of $z^{-1}$ functions is called the delay chain, as each $z^{-1}$ corresponds to a delay of one sampling period. The branches that come off

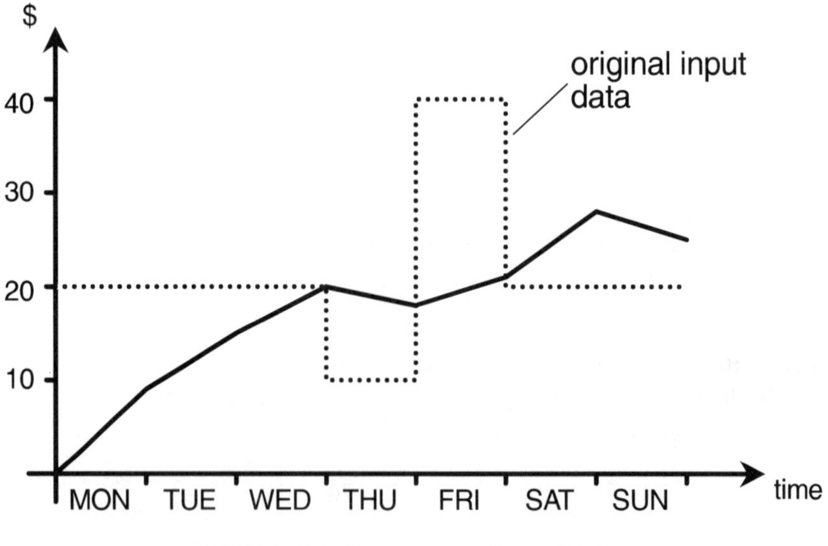


**FIGURE 4.22.** Output values from FIR filter

the delay chain are called taps, since they are the points at which we "tap-off" a certain proportion of the signal. The circuit in Figure 4.21 can therefore be described as a three-tap digital filter.

We still haven't explained why we call this a Finite Impulse Response filter. Let's go back to our calculation for Thursday's output. The value that represented the share price on Monday simply dropped off the end of our filter. This was entirely expected since we had said earlier that the output would only depend on the present sample and two previous ones. In other words, the output depends on a finite number of inputs, hence the name Finite Impulse Response.

Although we do not attempt to cover the mathematics of DSP in any detail in this book, it is useful to describe the various circuits mathematically so that all future reading is much simpler. In Chapter 3, we spent a little time reviewing the weighted impulse function. The digital filter's input signal $x(n)$ can be described as a series of these weighted impulses, as shown in Figure 4.23, and using these, Monday's input value can be written:

$$\int_{-\infty}^{+\infty} 20\,\delta(t) = 20$$

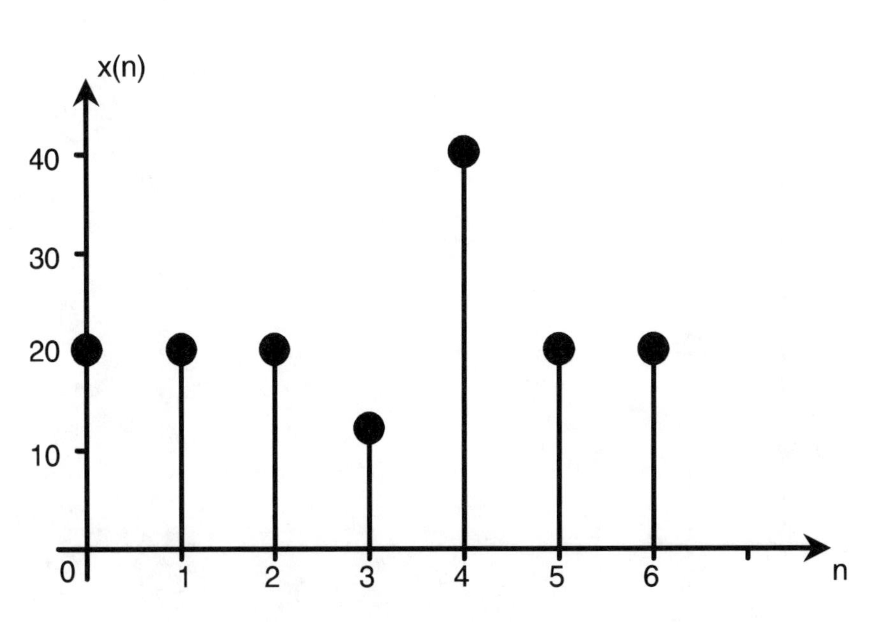


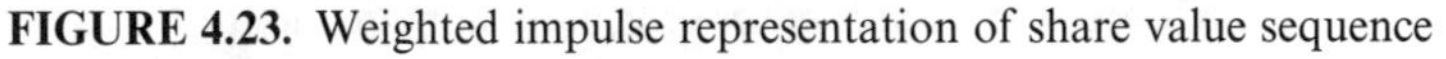
**FIGURE 4.23.** Weighted impulse representation of share value sequence

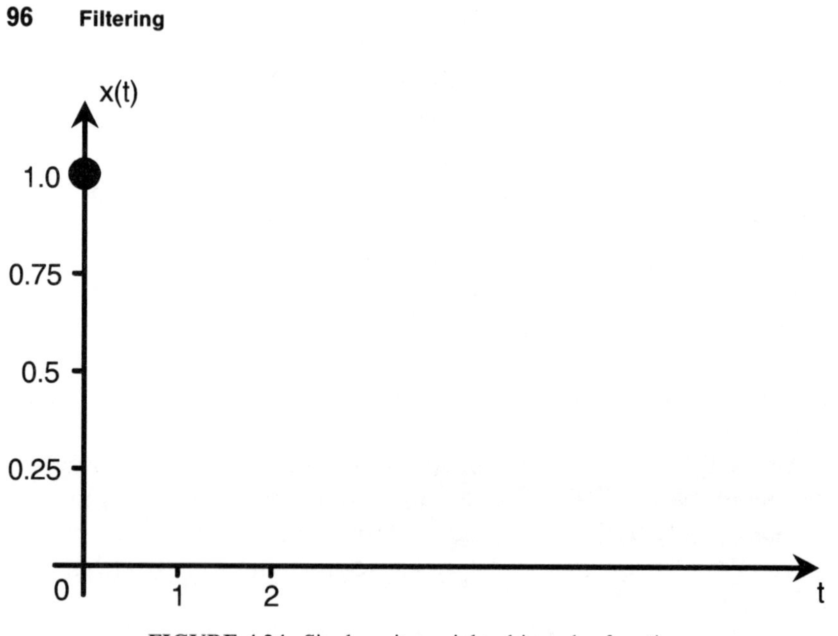


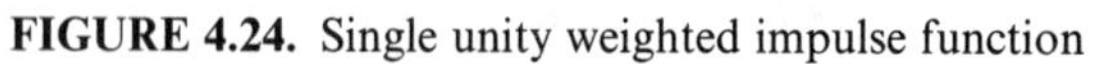

**FIGURE 4.24.** Single unity weighted impulse function

All the input values for the other days can be written in a similar manner. We can also use weighted impulse functions to describe the filter itself. The definition of a filter's impulse response is the output waveform when a single unity weighted impulse function is applied to

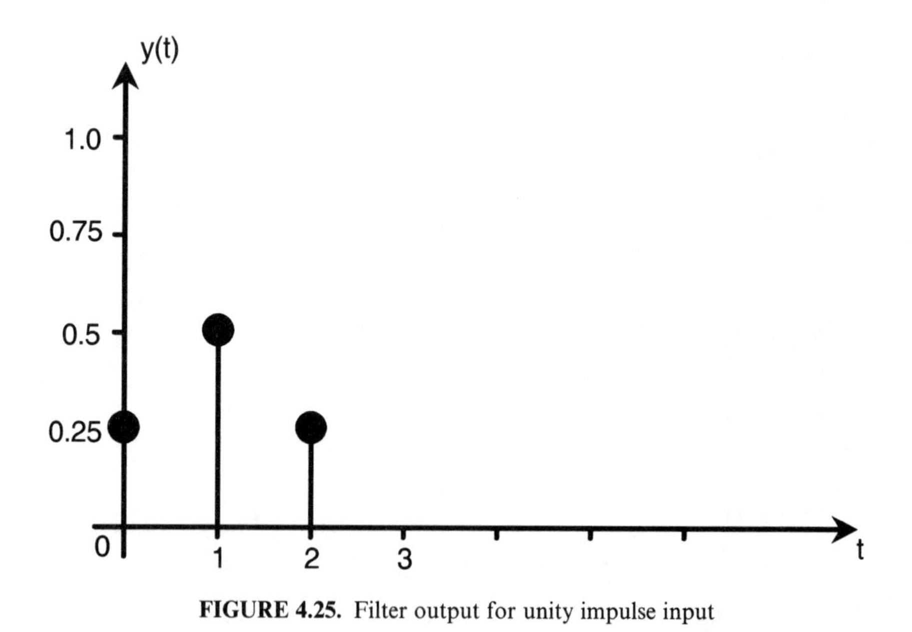


**FIGURE 4.25.** Filter output for unity impulse input

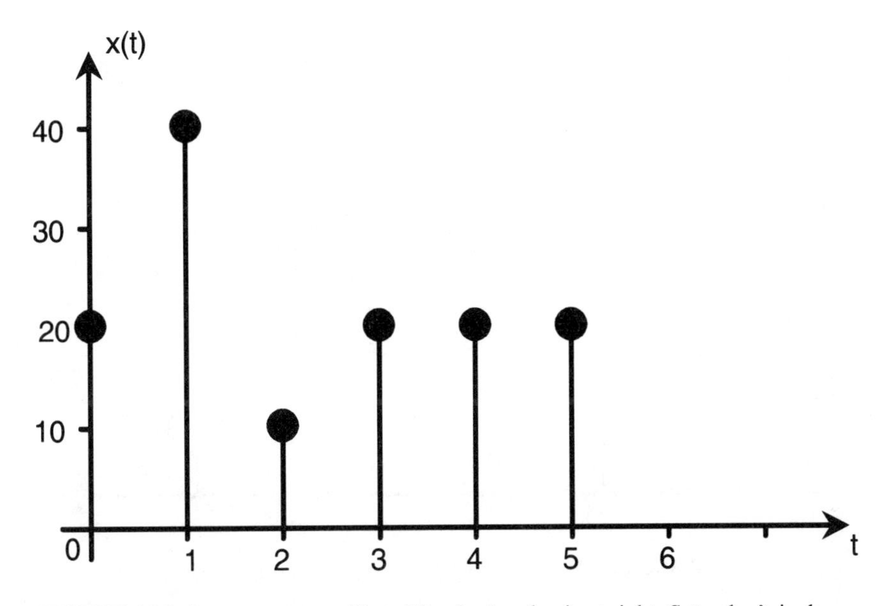


**FIGURE 4.26.** Input sequence. Note Monday's value is at right, Saturday's is shown at $t = 0$

the filter input at $t = 0$ (Figure 4.24). If we repeat the analysis that we performed above with the new single value input, we shall obtain the pulse train as shown in Figure 4.25.

Weighted impulse functions have the advantage that they allow us to describe the filter in a pictorial manner. It is then possible to calculate the output signal at any point in time simply by multiplying the impulse response of the filter by the input pulse train present at the desired time. Let us take just one instant, say $n = 5$; the input series $x(t)$ is shown in Figure 4.26. The present value (Saturday's) is shown at $t = 0$ and the previous values range *back* from that point (the value at $t = 5$ is Monday's). We can then see how easy this makes the calculation of the output for $y(5)$ – we simply multiply the values of $x(t)$ with the values of $a_R$ in the same position (Figure 4.27), giving:

$$y(5) = 0.25 \times 20 + 0.5 \times 40 + 0.25 \times 12 = 28$$

which is what we already worked out for Saturday's output. Moving on to Sunday, we simply shift the input impulse response along one stage to the right to allow Sunday's value to take up the position at $t = 0$ and redo the sums.

It is very interesting to note that to perform this we have in effect

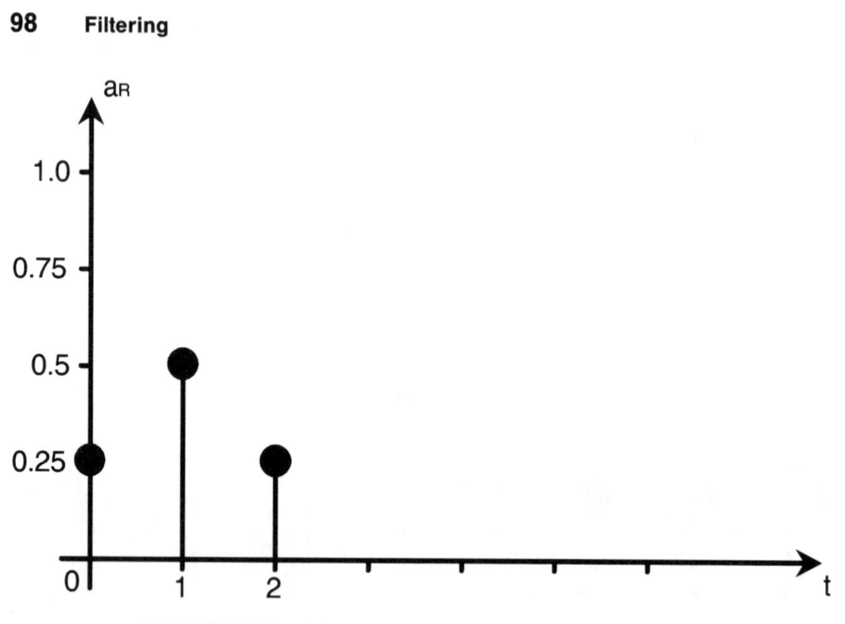


**FIGURE 4.27.** Filter coefficient values for example FIR

reversed the input sequence $x(n)$ before we multiplied it by the filter's impulse response. Then as we move on to calculate the next output, we simply shift the input sequence one position to the right and introduce the new value at $t = 0$. This technique of reversing the input and moving it past another set of impulse functions is usually referred to as convolution. We shall come back to this in a later chapter.

If you are a bit confused by this example remember that the sequence $x(t)$ is just our original sequence $x(n)$ drawn in reverse. In the case of the sequences shown in Figures 4.23 and 4.26, $t = 5 - n$, so when $n = 5$ (Saturday), $t = 0$ and when $n = 4$, $t = 1$, etc.

We initially described the filter using the following equation:

$$y(n) = a_0x(n) + a_1x(n-1) + a_2x(n-2)$$

We now need to develop our filter description a little further. We have said that the unit delay is defined as $z^{-1}$. From this we define the delayed input samples as:

$$\begin{aligned} x(n-1) &= x(n) \cdot z^{-1} \\ x(n-2) &= x(n) \cdot z^{-1} \cdot z^{-1} \\ &= x(n) \cdot z^{-2} \end{aligned}$$

So we can rewrite the equation for the filter as:

$$y(n) = (a_0 + a_1 z^{-1} + a_2 z^{-2}) \cdot x(n)$$

Alternatively:

$$H(n) = \frac{y(n)}{x(n)} = a_0 + a_1 z^{-1} + a_2 z^{-2}$$

The function $H(n)$ is the mathematical representation of the impulse response of the filter. More often it is called the transfer function (similar to $H(\omega)$ in analog filters at the start of the chapter). A simple rule to remember is that if $z^{-n}$ (where $n$ is any number) appears only on the numerator of the transfer function, that filter is inherently stable. Unlike the average analog design, this filter will always be a filter – never an oscillator!

One of the most significant features of the FIR filter is that it can be adapted to construct a linear phase response. To do this we simply mirror the values of the coefficients $a_R$ around the center tap, so that:

$$a_0 = a_R$$

$$a_1 = a_{R-1}$$

etc.

where $R + 1$ is the number of filter taps.

### Infinite Impulse Response Filters

The other basic form of a digital filter is called the Infinite Impulse Response (IIR) filter. A simple form of this is shown in Figure 4.28. Using the same notations as we have just used for the FIR, we can see that:

$$\begin{aligned} y(n) &= x(n) + a_1 y(n-1) + a_2 y(n-2) \\ &= x(n) + [a_1 z^{-1} + a_2 z^{-2}] \cdot y(n) \\ &= x(n) \frac{1}{1 - a_1 z^{-1} - a_2 z^{-2}} \end{aligned}$$

Take the math for granted – it's just relatively simple substitution.

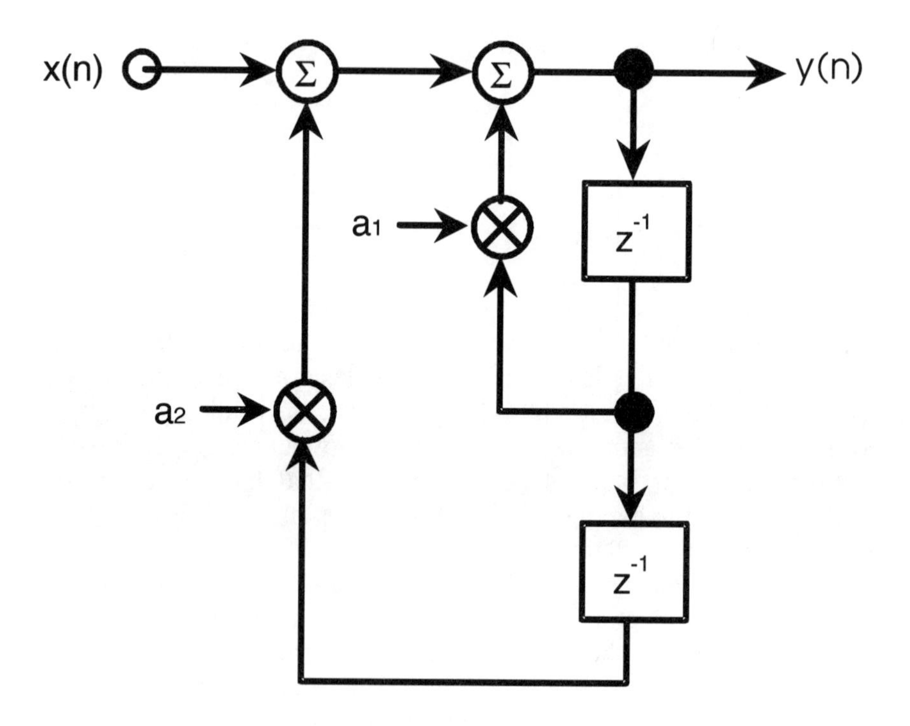


**FIGURE 4.28.** A simple Infinite Impulse Response Filter

Therefore, the transfer function is given by:

$$H(n) = \frac{y(n)}{x(n)} = \frac{1}{1 - a_1 z^{-1} - a_2 z^{-2}}$$

From the first equation we can see that each output, $y(n)$, is dependent on the input value $x(n)$ and two previous outputs, $y(n-1)$ and $y(n-2)$. Taking this one step at a time, let us assume that there were no previous input samples before $n = 0$. Then:

$$y(0) = x(0)$$

At the next sample instant:

$$\begin{aligned} y(1) &= x(1) + a_1 y(0) \\ &= x(1) + a_1 x(0) \end{aligned}$$

Then at $n = 2$:

$$y(2) = x(2) + a_1 y(1) + a_2 y(0)$$
$$= x(2) + a_1[x(1) + a_1 x(0)] + a_2 x(0)$$

And at $n = 3$:

$$y(3) = x(3) + a_1 y(2) + a_2 y(1)$$
$$= x(3) + a_1[x(2) + a_1[x(1) + a_1 x(0)] + a_2 x(0)]$$
$$+ a_2[x(1) + a_1 x(0)]$$

We can already see that any output is dependent on all the previous inputs and we could go on, but the equation just keeps getting longer. An alternative way of expressing this is to say that each output is dependent on an infinite number of inputs. This is why this filter type is called Infinite Impulse Response (IIR).

If we look again at Figure 4.28, the filter is actually a series of feedback loops and as with any such design, we know that under certain conditions it may become unstable. Although instability is possible with an IIR design, they have the advantage that for the same

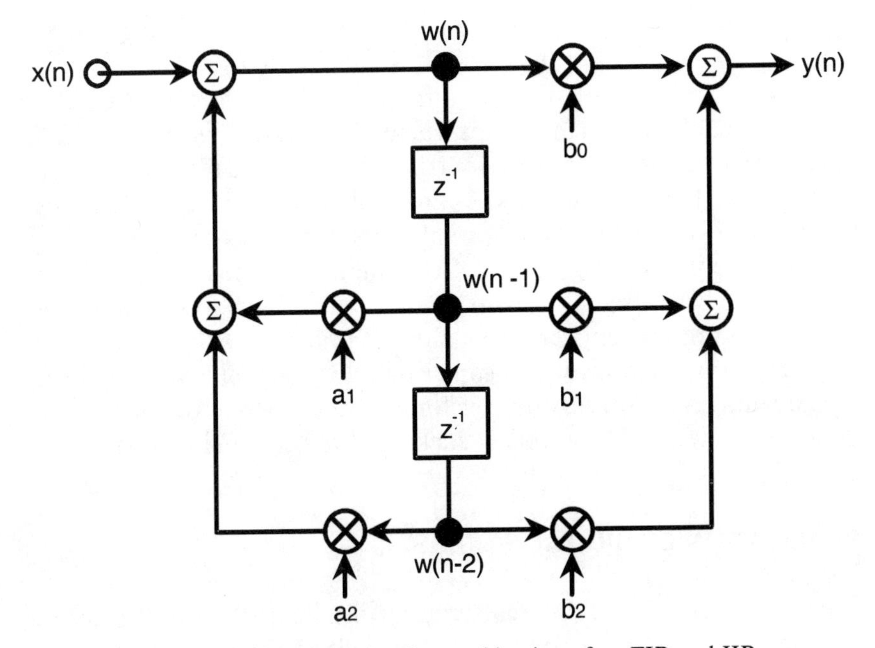


**FIGURE 4.29.** A second-order combination of an FIR and IIR

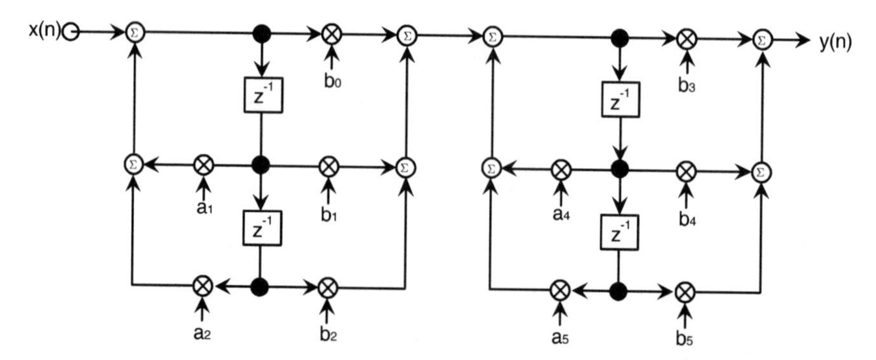


**FIGURE 4.30.** A fourth-order 'cascaded' filter using two second order structures

roll-off rate they require less taps than FIR filters. This means that if we are limited in the processor resources available to perform our desired function, we may find ourselves having to use an IIR. We just have to be careful to design a stable filter.

IIR filters are commonly used for ac coupling and smoothing (averaging) but it is more usual to see combinations of FIR and IIR structures. For example, the direct form as shown in Figure 4.29 is very common. The operation of this filter is given by:

$$y(n) = (b_0 + b_1 z^{-1} + b_2 z^{-2}) \cdot w(n)$$

$$w(n) = x(n) + (a_1 z^{-1} + a_2 z^{-2}) \cdot w(n)$$

If we require filters with more taps we simply cascade this second-order (two-tap) form. For example, a fourth-order cascaded filter is shown in Figure 4.30. The advantage of designing filters in this way is that they are modular, which makes their implementation much simpler.

There are many other forms of IIR filter. The lattice filter is quite common. There are also many ways to combine a second-order sections to produce higher-order filters. We have looked only at the cascaded form; obviously they can also be combined in parallel. There are advantages and disadvantages for all these different designs. For a detailed analysis of filters look at Jackson [1989] and many other texts.

## REALIZATION OF DIGITAL FILTERS

We introduced several different types of filter earlier in this chapter, for example, the Butterworth and Bessel filters. Each of these styles optimized a specific characteristic in either the gain or phase response of the filter. These filter types are found in both analog and digital designs and transformation techniques are used to convert the analog circuit descriptions into digital filters.

In any further reading you do, you will rarely find any digital approximations to the Bessel function, since as we have already seen it is possible to design a linear phase filter simply by using an FIR circuit.

These transformation techniques do not help us understand digital filtering – they are very mathematically intensive and as such are not covered in this book. Bateman and Yates [1988] or again Jackson [1989] go into this subject in considerable depth.

The question still remains, however: What makes these digital filters easy to implement on a processor and how does a DSP do it so much better than any other type of processor? We can begin to explain this if we go back to our original block diagram of a filter, as shown in Figure 4.31. We will use this second-order FIR filter to clarify the explanation.

The equation for the output is given by:

$$y(n) = a_0 x(n) + a_1 x(n-1) + a_2 x(n-2)$$

Alternatively, for any order FIR we can say:

$$y(n) = \sum_{R=0}^{M} a_R x(n-R)$$

where $M$ is the order of the filter.

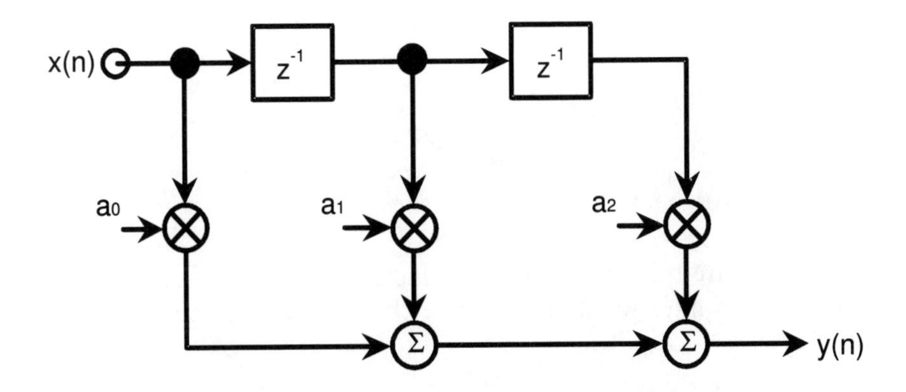


**FIGURE 4.31.** Basic second-order FIR filter

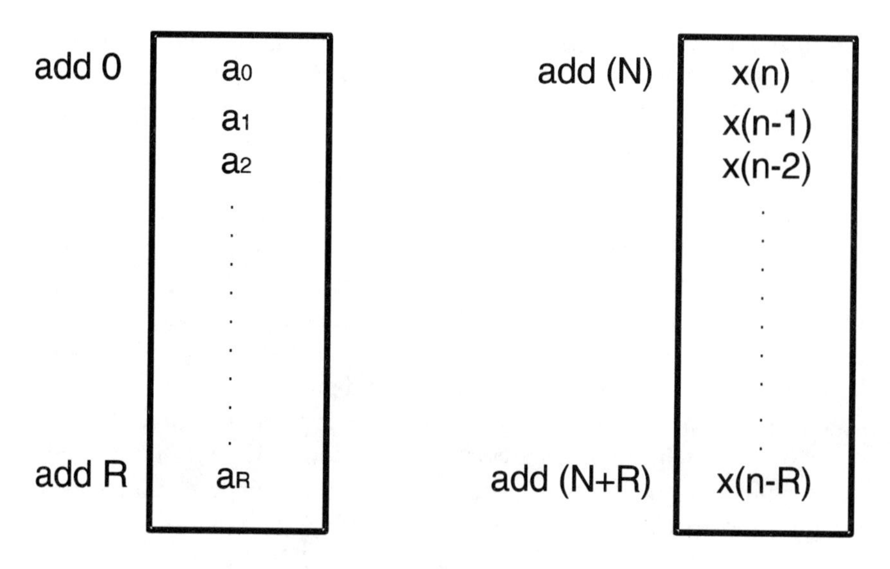


**FIGURE 4.32.** Tables of coefficients ($a_R$) and data ($x(n-R)$)

The filter is composed of a set of coefficients $a_R$ and a set of input values $x(n)$, which we can represent as tables as shown in Figure 4.32. In order to compute the output at any one time we then simply have to multiply the corresponding values in each table and sum the results. If we were to write the required function for a three-tap (second-order) filter in longhand we should say:

subroutine FIR2:

input $x(n)$

$i(0) = a_0 x(n)$

$i(1) = a_1 x(n-1)$

$i(2) = a_2 x(n-2)$

$y(n) = i(0) + i(1) + i(2)$

where $i(M)$ are the intermediate products. Thus, we can see that for our three-tap filter we should require three multiplications and two additions:

$$y(n) = a_0 x(n) + a_1 x(n-1) + a_2 x(n-2)$$

General-purpose microprocessors take many cycles to compute a multiplication. This is because they perform a long multiplication:

$$
\begin{array}{r}
A_1 A_0 \\
\underline{B_1 B_0} \\
B_0 A_1 + B_0 A_0 \\
B_1 A_1 + B_1 A_0 + 0 \quad \\
\hline
C_2 \quad + C_1 \quad + C_0 \quad
\end{array}
$$

Let us briefly take a two-bit binary multiplication as an example:

$$
\begin{array}{r}
10 \\
\underline{11} \\
10 \\
\underline{100} \\
0110
\end{array}
$$

We can extend the example to show that to multiply two 16-bit numbers we must compute 256 binary multiplications, perform 15 left shifts and then do 32 binary additions of the partial results. This appears to be a formidable task; however, most general-purpose microprocessors can perform all these functions in approximately 10 machine cycles, i.e. a 20MHz processor can perform a 16-bit multiply in around $0.5\mu s$. Taking this example to the logical conclusion, we can see that to calculate an output from a two-tap filter will take the 20MHz processor around $1.05\mu s$ if we assume that it can perform a 16-bit addition in a single cycle.

On the other hand, general-purpose DSPs have the multiplication function designed into their silicon, which means that they can perform the multiplication of two $n$-bit numbers in a single cycle. The TMS320 family of DSPs go further by being capable of performing an $n$-bit multiplication plus a $2n$-bit addition in one machine cycle. The reason for the addition being $2n$ bits can be easily understood by looking at the 2-bit example above, i.e., if we multiply two 2-bit numbers then the maximum possible result will be 4 bits long $(11 \times 11 = 1001)$.

To illustrate the advantage of having a hardware multiplier, let's examine the performance of the TMS320C5x DSPs, which have a machine cycle time of 25ns. With these DSPs we can perform the mathematics for our two tap filter in just 3 cycles, or $75\text{ns} = 0.075\mu s$.

This is a very important difference between standard microprocessors and DSPs. If we extend the analysis a little further, we can take a realistic filter of 50 taps. Our DSP could perform the mathematics in around 1.275$\mu$s, meaning that it can accept a new input sample every 1.275$\mu$s. This implies that it could handle a sampling frequency of 0.78MHz. Going back to the sampling theorem ($f_s \geq 2f_m$), the filter is therefore capable of handling input signals at frequencies of up to 392kHz.

In comparison, a general-purpose microprocessor with a comparable speed of 33MHz (machine cycle time = 30ns) takes 10 cycles for each multiply and one cycle for an addition. If we implemented a similar 50-tap filter, it would only be able to accept input signals with frequencies up to 30kHz.

Pure multiplication speed is only a part of the calculation of the performance of different processors for any application. By itself, it does not give us sufficient information on the real ability of a device to calculate digital filters. Looking back at the examples of filters we have previously used, we can see that in general there are long lists of data that must be multiplied in the correct order. Additionally, although the coefficient values are static, the input data changes every sample period, e.g., the $x(n)$ value for one sampling period becomes $x(n-1)$ in the next, then $x(n-2)$, etc., until it simply drops off the end of the delay chain.

The most efficient method for handling these data tables is to load all the input values into a circular buffer (Figure 4.33). The beginning of the data, $x(n)$, is located by a pointer as shown and the previous data values are loaded sequentially from that point in a clockwise direction. As we receive a new sample it is placed at the position $x(n)$ and our filter subroutine starts at $x(n)$, multiplying the data values with the corresponding tap value, $a_R$. After calculating the final output, the pointer is moved counterclockwise one position to point at $x(n-R)$ and we wait for the next input sample. The next input is written to the $x(n-R)$ position, which is permitted because the old $x(n-R)$ data dropped off the end of our delay chain after the previous calculation. We then simply repeat the filter subroutine as before.

This circular buffer sounds ideal: it saves moving all the data each time we receive a new input sample. But it is too good to be true unfortunately. Take a look at real microprocessors – how many have you seen with circular memory spaces of variable length?

All types of microprocessor, including digital signal processors,

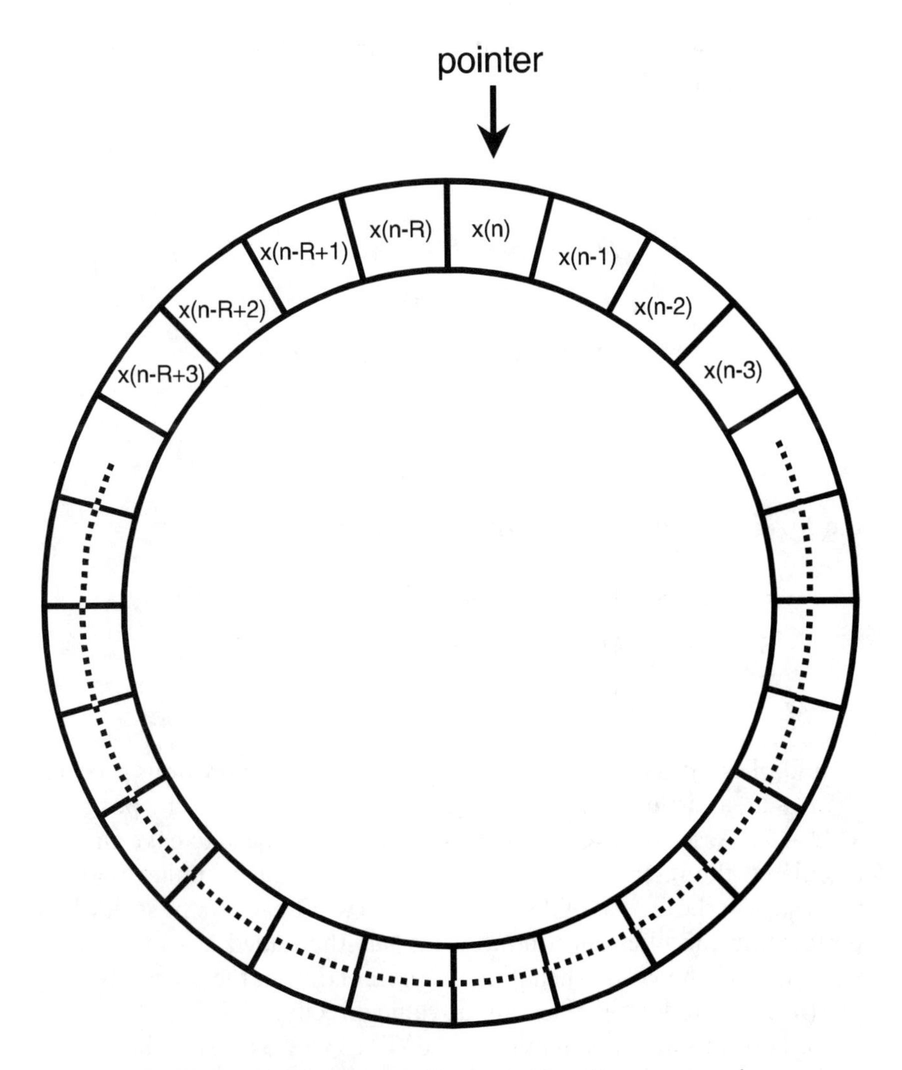


**FIGURE 4.33.** A circular buffer for holding filter data samples

have linear memory spaces. By that we mean that the lowest address is 0, the highest XXXX (hex) and if you try and move a value past the bottom of the memory it will either cause an error or the information will just be lost. So how do we get around this?

There are two main methods used in DSP: linear addressing with data moves or simulated circular buffers. Many DSPs are organized such that data moves can be performed in parallel with the multiply accumulate (MAC) function. For example, the TMS320C2x has the instruction "MACD," which is performed in a single cycle. Taking our general filter requirement, if we are in the middle of our calculation, say at tap $M$:

$$\begin{aligned}
&\vdots\\
y(M-1) &= i(M-1) + y(M-2)\\
i(M) &= a_M x(M)\\
y(M) &= i(M) + y(M-1)\\
&\vdots
\end{aligned}$$

the MACD instruction performs the following:

$$\begin{aligned}
i(M) &= a_M x(M)\\
y(M-1) &= i(M-1) + y(M-2)\\
x(M) &:= x(M-1)
\end{aligned}$$

It multiplies the data and coefficient, accumulates the previous product ($i(M-1)$) and moves the data value $x(M)$ to the previous data position, $x(M-1)$. As MACD writes $x(M)$ to $x(M-1)$ we need to have already calculated the intermediate product $i(M-1)$, which implies that we must start at the far end of the delay line and work back to the start when performing the filter subroutine. This has the added advantage that when we have finished calculating the present output the pointer is back at $x(n)$ ready to load the next input value directly.

In addition to this method, the TMS320C5x also allows the designer to implement two simulated circular buffers using a number of special registers. There are separate start of buffer (SOB) and end of buffer (EOB) registers plus two control registers, including flags to indicate full or empty buffers. Once the registers are correctly programmed, the memory of the TMS320C5x behaves like a circular buffer and all supervision of where the data is stored to/retrieved from is controlled by the DSP itself and is transparent to the user.

Both of the above techniques are perfectly acceptable to any designer. The only advantage of the second method is that it offers a simpler solution to people who do not want to understand the architecture of the device they are using.

Summarizing, we have seen that we use DSPs to implement FIR filter designs because they can perform the mathematics much faster than any general-purpose microprocessor. It is easy to extend the argument to IIR filters, which are also just a combination of tables to be multiplied and accumulated.

## COMB FILTERS

We shall take a little time here to look at a special type of filter that is found in the output stage of sigma delta ADCs. As we saw in Chapter 3, these ADCs use output decimation filters first to convert the high-speed single-bit data stream into a lower-speed multibit value and then to remove any noise pushed back into the frequency range of interest by the crude 1-bit modulator.

The characteristic shape of a comb filter is shown in Figure 4.34 and its gain response is given by:

$$H(z) = \frac{1 - z^{-k}}{1 - Az^{-k}}$$

Figure 4.35 shows a block diagram of the comb filter that can be implemented on our DSP using the following subroutine:

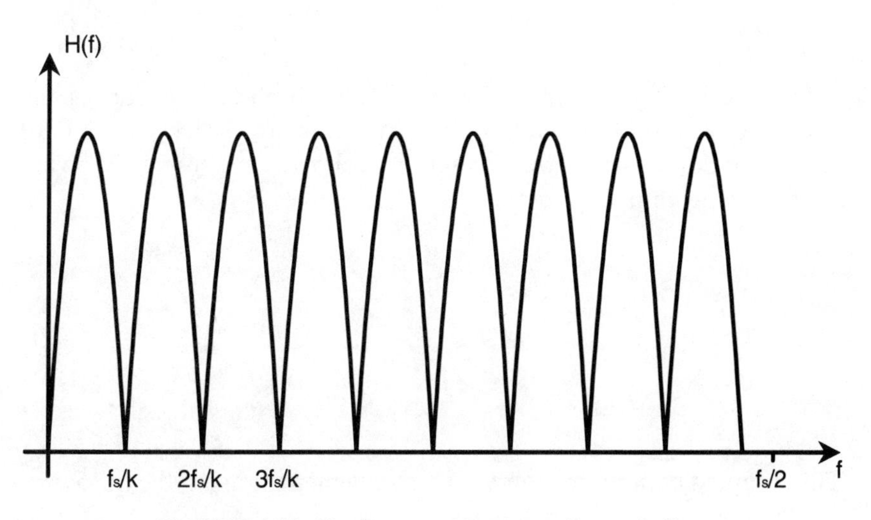


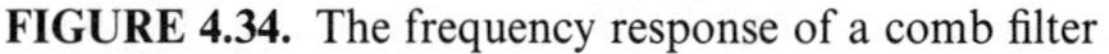
**FIGURE 4.34.** The frequency response of a comb filter

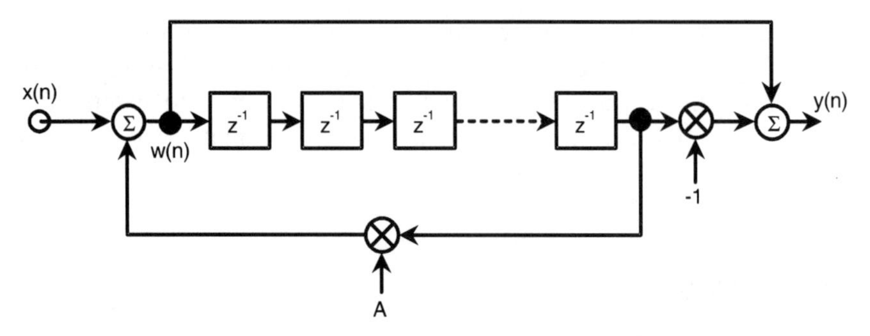


FIGURE 4.35. Block diagram of a comb filter

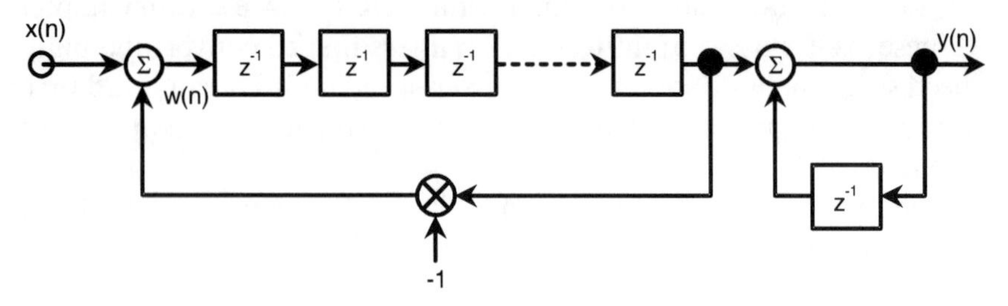


FIGURE 4.36. Modified form of comb filter for implementation on a DSP

subroutine comb:

$$i(0) = A \cdot w(n-k)$$
$$w(n) = i(0) + x(n)$$
$$y(n) = w(n) - w(n-k)$$

In the case of the sigma delta ADC, our comb filter is part of the device, so we want to build it in silicon rather than use a separate DSP device. To do this we modify it slightly as shown in Figure 4.36, giving the following gain response:

$$H(z) = \frac{1 - z^{-k}}{1 - z^{-1}}$$

$$= \frac{1}{1 - z^{-1}}(1 - z^{-k})$$

This adapted comb filter can then be implemented using the following subroutine:

subroutine adcomb:

$$i(0) = x(n) - x(n-k)$$
$$y(n) = i(0) + y(n-1)$$

The adapted comb filter now requires only addition and subtraction. Therefore it uses a very small area of silicon and is relatively cheap to build.

The comb filter is only part of the output circuit of the sigma delta ADC. It is followed by an FIR decimating filter that averages the single-bit output. Let's assume that the decimator is of order $q$, i.e., it converts the single-bit input sample to a signal at $1/q$th of the input frequency and $k$ bits wide (Figure 4.37). We can rearrange the different elements of the output so that the decimator is in the middle, reducing the differentiation to a single stage as in Figure 4.38 (see references for a full mathematical treatment).

The components of the comb filter are now simply an integrator and a differentiator:

$$\text{integrator}: \quad y(n) = x(n) + y(n-1)$$
$$\text{differentiator}: \quad y(n) = x(n) - x(n-1)$$

which are again easy to build in silicon. In general, cascaded comb filters are used to increase the attenuation in the stop band to an acceptable level. The frequency responses of first and fourth-order

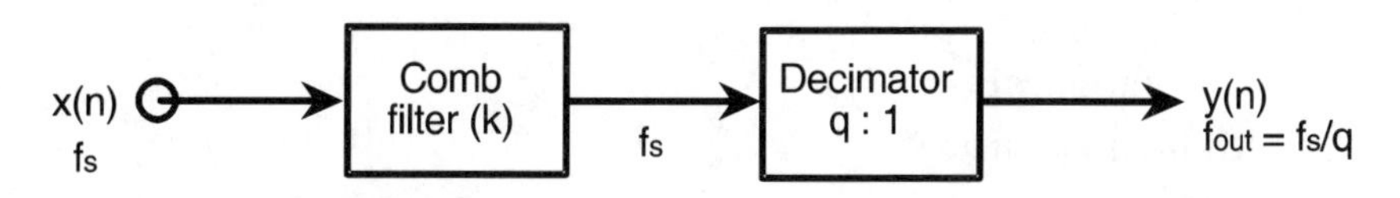


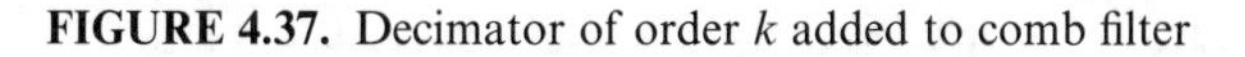

**FIGURE 4.37.** Decimator of order $k$ added to comb filter

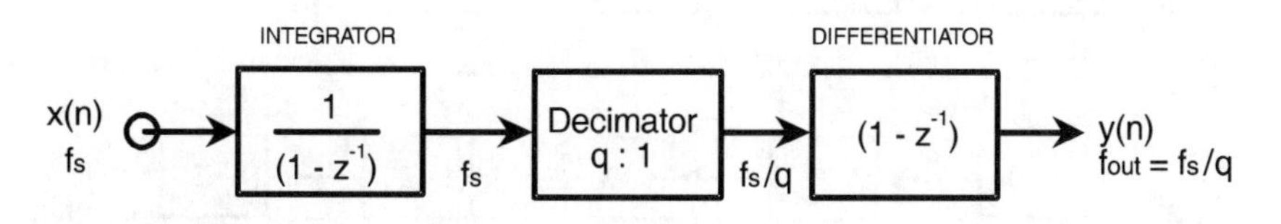


**FIGURE 4.38.** Rearranged output circuit of sigma delta ADC, with comb filter divided into integrator and differentiator

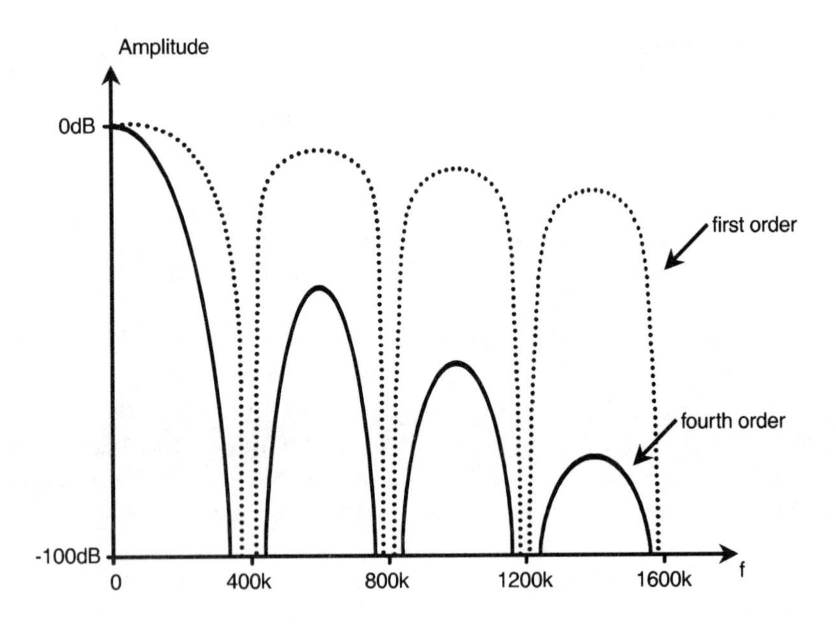


**FIGURE 4.39.** Frequency response of a first- and fourth-order comb filter

comb filters are shown in Figure 4.39 with the block diagram of the fourth-order comb filter shown in Figure 4.40.

The log amplitude response of a sixteenth-order comb filter with a 16:1 decimator is shown in Figure 4.41. We can see that the disadvantage with this type of filter is the significant amount of "droop" in the gain response in the first lobe, which is our pass band. For this reason, sigma delta ADCs usually include a "compensation" FIR filter in the final stage.

In summary the comb filter has the following advantages:

1. No multiplication
2. No filter coefficients

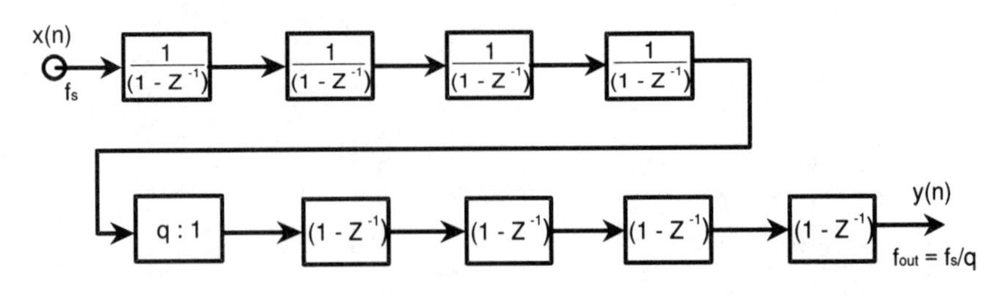


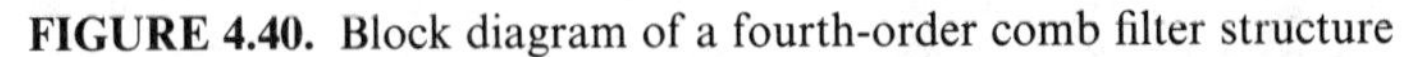
**FIGURE 4.40.** Block diagram of a fourth-order comb filter structure

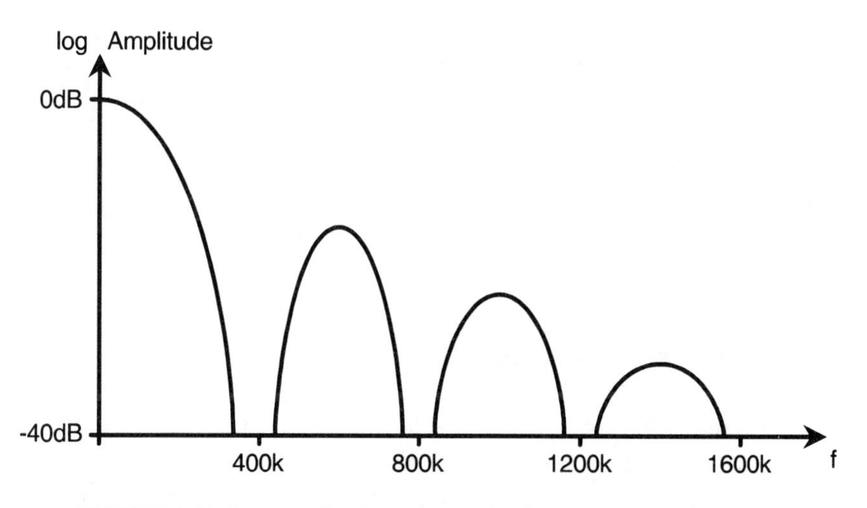


**FIGURE 4.41.** Log amplitude response of a sixteenth-order comb filter

3. Simple to extend to higher order
4. Easy to design – the value $k$ defines filter notches
5. Can be used for much higher sampling rates as the filter chain is short when compared with the equivalent multitap FIR filter.

## COMPARISON OF FIR AND IIR FILTERS

FIR filters are generally preferred where the phase response of the circuit is of primary importance. They can be designed with precise linear phase and as they are stable and predictable in nature, they are preferable if we require a high level of control over the phase response of our system.

In contrast, IIR filters usually have very poor phase responses that are extremely nonlinear at the edge of the bands. Some software filter design programs completely ignore the IIR filter's phase response as a design parameter. The reason is that with IIR filters we are usually looking for the best roll-off for the smallest number of taps and we are prepared to sacrifice the phase response. For a given number of taps the IIR filter is five to ten times more efficient in controlling its gain response than the FIR filter.

This comparison allows us to examine the economics of implementing filters in terms of hardware complexity and computational speed. If we are limited in either of these two areas it may be that we need to use an IIR design to achieve our specified filter response. The main differentiating feature between the two basic types of filter is that a pure FIR filter, i.e., one with feed-forward elements only, is inherently stable. In comparison, the IIR filter always contains some feed-back components and, as we know, these may cause the circuit to become unstable under certain conditions. Therefore and the filter must be carefully designed.

## NOISE IN FILTER DESIGNS

As we saw in Chapter 3, a design that has been digitized has inherent errors as the circuit can only be an approximation to the original analog system. In general, the effects of the noise introduced due to these errors is more pronounced in IIR filters, because their feed-back elements cause the errors to be accumulated over time. FIR filters, however, are simply feed-forward circuits so that any errors appear in the output signal only once per sampling period.

We shall examine the five main types of error that are found in digital filter designs and the effects that they have on the performance of the circuit. Further information on this topic can be found in Jackson [1989].

### Signal Quantization

We know that all analog-to-digital converters (ADCs) introduce quantization errors to the incoming signal. In the case of a successive approximation or dual slope, the noise introduced is directly proportional to the number of bits in the output. The sigma delta ADC is slightly different. The noise it introduces is dependent on the design of its modulator and filters.

In addition, when the output digital-to-analog converter (DAC) resolution is less than the internal resolution of the DSP, it also adds a certain amount of noise to the output signal. However, it should be noted that this noise is not circulated through the filter and so cannot be influenced by its design.

If we use a 10- or 12-bit ADC/DAC combination with a 16-bit DSP, the maximum noise possible at the input to the filter is $2^{-10}$ or $2^{-12}$. If we assume that the filter has a gain of less than 0dB over its whole frequency range, the noise added by these sources is negligible.

### Coefficient Quantization

The effects of coefficient quantization are most significant in fixed-point DSPs. Larger floating-point DSPs of 32 bits or more are sufficiently accurate that these effects can be ignored. Fixed-point DSPs generally have word lengths of 16 bits, and all data is expressed as fractions in order to avoid overflow in the DSP's accumulator. The accuracy of the coefficients is important, as we shall cover in detail in Chapter 7, because in multiplying two fixed-point numbers together we may amplify any error in the values. For example, let us multiply an input of 0.39 by 0.4:

$$\begin{array}{l} 0.39 \\ 0.4 \\ \hline 0.156 \end{array}$$

An error of one decimal bit in the coefficient would change 0.4 to 0.3 and then the result would become:

$$\begin{array}{l} 0.39 \\ 0.3 \\ \hline 0.117 \end{array}$$

which is in error by 25%. In general, 16-bit coefficients are adequate to ensure such errors are insignificant. An error in the least significant bit (LSB) is then $<0.002\%$.

In the case of IIR filters an error in the LSB of a coefficient may be much more disastrous. A single LSB error in one feed-back element may cause only a small error on the first pass, but it may result in a very significant error after many loops around the filter and cause it to become unstable.

Most software filter design packages also allow the user to examine the stability of IIR filters. Before committing a filter to silicon it is worthwhile using this facility to ensure that any small variations in the coefficients will not cause problems in the final design.

### Truncation Noise

If we multiply two $n$-bit numbers together we require $2n$ bits to store the answer. This can again be demonstrated using a decimal example:

```
  0.64
  0.73
  ----
   192
  4480
------
0.4672
```

We can show that this is also true for binary numbers:

```
  11
  11
 ---
  11
 110
----
1001
```

This is why in all fixed-point DSPs the product register and the accumulator are double the width of all other registers. For example, in the 16-bit TMS320C25 the product ($P$) register and the accumulator (ACC) are both 32 bits wide.

The advantage of this limited extra precision during the calculation of the filter output can be seen if we examine the following pseudocode for the filter loop, where $P$ denotes the $P$-register and $A$ the accumulator:

subroutine filter:

.

.

$P = a_{R-1}x(n - (R - 1))$

$A = A + P$

$P = a_R x(n - R)$

$A = A + P$

$P = a_{R+1}x(n - (R + 1))$

$A = A + P$

.

.

As each 32-bit product is added to the accumulator, the 32 bits of precision are maintained during the filter subroutine. The 32-bit result must then be stored in 16-bit-wide memory. It is possible to use two instructions to store both halves of the result in consecutive memory locations, but this doubles the time taken to store the output and the amount of memory required. It also doubles the time required to retrieve the value for use in subsequent calculations. Because of this overhead it is usual to store only the most significant 16 bits and truncate the result.

The error due to this truncation is only in the 16th bit and represents $<0.001\%$. There is also the option in the TMS320Cxx range of DSPs to round the result, which reduces the maximum possible error even further. In the majority of real-world applications this provides more than enough precision.

## Internal Overflow

In most processors, when a mathematical function produces an output that is greater than the processor's accumulator, the result simply "wraps round". For example, if we add 0110 to 1111, the result will be 10101. In a 4-bit processor only the lower four bits can be held in the accumulator so that the MSB is lost. The answer to our addition is then 0101, which is obviously incorrect. Similarly, underflow also causes the processor to wrap round in the negative direction.

As with all the other error sources previously discussed the effect of internal overflow, or underflow, is not catastrophic in FIR filters. This is because the error will be confined to a single output. Conversely, with IIR filters an error in one output will be fed back into the filter and cause all subsequent outputs to be in error.

In order to avoid these effects, DSPs allow the designer to switch to a special mode called *saturation*. In this mode, if any result is greater than the largest possible value, the DSP's accumulator will saturate to all ones and if the result is less than the smallest acceptable value, the output will saturate to all zeros. This ensures that incorrect values are not circulated in the filter, but it is clearly not an ideal situation: if the DSP saturates when performing the filtering algorithm it will not achieve the desired result.

We should try to design the filter to avoid such situations. Manufacturers of DSPs always provide software simulator programs that we can use to simulate filter operation. We can check for any occurrence of internal overflow and simply scale the input signal to remove any problems.

### Dynamic Range Constraints

The dynamic range of a DSP may limit the performance of the filter as it is directly related to the word width of the processor. The number of available bits is related to the minimum size of quantization step available and therefore the granularity of the filter. As we saw in Chapter 3, if a device has a limited number of bits we need to use a more sophisticated quantization scheme to accurately capture all the meaningful information in the incoming signal.

The fixed-point range of the TMS320Cxx DSPs all have word widths of 16 bits. This gives a dynamic range of:

$$\begin{aligned} 20\log_{10}(2^{16}) &= 20\log_{10}(65536) \\ &= 96\text{dB} \end{aligned}$$

In the arithmetic unit, the accumulator and product register have been extended to 32 bits, providing a dynamic range of 192dB for the filter calculation. This is more than adequate for most filtering applications.

## FILTER DESIGN PACKAGES

We have not at any time discussed how to determine the actual values of the coefficients required to build a digital filter. As we stated earlier, this is not an easy task and is generally best performed using a computer software package.

Using these packages it is possible to choose a filter design quickly and easily, calculate the filter coefficients and simulate the filter response in both gain and phase. If our filter is then not quite what we wanted, the process can simply be repeated. There are several software packages that also produce TMS320Cxx assembly code for the filter design – making the software design extremely simple.

We shall quickly look at one such package called DFDP (Digital Filter Design Package) from Atlanta Signal Processors Inc. which runs on any IBM-compatible PC. This software is capable of designing FIR filters with up to 510 taps.

The program requires the user first to specify the sampling

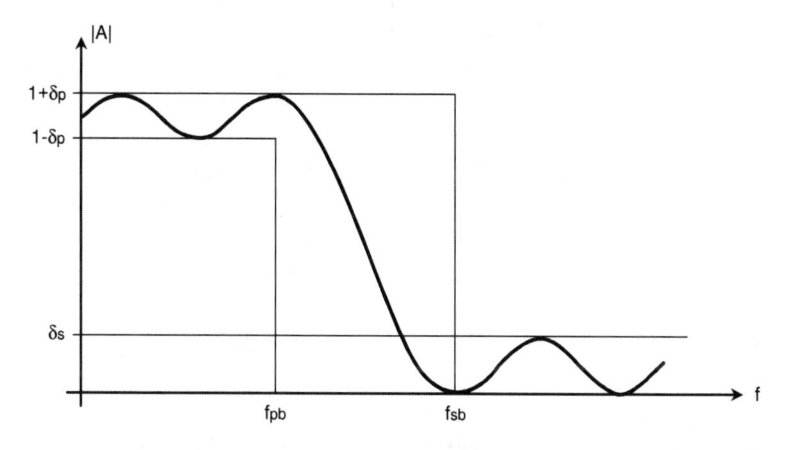


**FIGURE 4.42.** Filter response specified in digital filter design package

frequency to be used. All subsequent frequencies that are given to the program must be $\leq {}^{1}/_{2}\, f_s$ to satisfy the Nyquist criterion. The filter response is then specified as shown in Figure 4.42.

The transition region is defined using two frequencies, $f_{pb}$ and $f_{sb}$. The roll-off rate in this region for any filter type is directly related to the number of taps in the design. Therefore, the faster we design the roll-off the more taps the software will compute. Note that it would take an infinite number of filter taps to design a filter with a "brick-wall" characteristic.

The allowable ripple in the pass and stop bands is defined by $\delta_p$ and $\delta_s$, respectively, where the pass-band ripple, $R_{pb}$, is given by:

$$R_{pb} = 20 \log_{10} \frac{(1+\delta_p)}{(1-\delta_p)}$$

and the stop-band ripple, $R_{sb}$, is given by:

$$R_{sb} = 20 \log_{10} \delta_s$$

Figure 4.42 shows a low-pass filter but the program can also compute the coefficients for all the other types: high-pass, band-stop, band-pass and multiband filters. The characteristics of all of these other filters are entered in the same format as shown for the low-pass filter.

The program first computes how many taps it believes is necessary to implement the filter and we can accept this or enter our own number.

There are two main reasons why we may wish to change the number of taps. First, if the filter is too long we may decide to trade the filter performance for fewer taps. Second, we may wish to force the program to alter the number of taps by $\pm 1$ because if we force an FIR filter to have an odd number of taps, we can produce a linear phase filter.

Once we accept the number of taps the software computes the coefficients and plots the corresponding frequency response. We are offered the option to view and plot the performance in several different ways including gain response, log gain response, impulse response and phase response.

If we decide that the filter does not actually meet our requirements in any particular way, we can simply restart the design loop using slightly adjusted figures. As we mentioned earlier, once the desired response is achieved we can directly produce TMS320Cxx assembly code using the DFDP.

## ADAPTIVE FILTERS

Throughout this chapter we have discussed various types of digital filters that were all time invariant. This means that their coefficients remained fixed throughout the life of the filter. An adaptive filter has coefficients that are designed to vary with time in a tightly controlled manner and it is usually assumed that this happens without the intervention of a human.

While not necessarily required, we also assume that the frequency of this variation in coefficient values is much lower than the bandwidth of the signal being filtered. In general, we use adaptive filtering techniques when we do not fully understand the properties of the signal we wish to filter. This may be due to genuine uncertainty about the incoming signal or because it will change slightly with time. In this way our adaptive filter will be able to change with the signal, or to "learn" the incoming signal's characteristics.

An example of the need for adaptive filters is the use of very precise sensors, especially where the sensors are connected to the monitoring unit by a long length of cable. As the equipment will probably be powered from the mains, the cable will suffer from 60Hz (or 50Hz) hum. The obvious answer would appear to be simply to high-pass filter the output from the sensor above say 100Hz, but there are many applications that cannot tolerate the total exclusion of all signals below 100Hz, for example in an electrocardiograph.

We could use a very narrow band-stop or notch filter to avoid removing too much of the incoming signal, but would our interfering signal then stay at a constant frequency? Probably not. We need to build an adaptive filter that monitors the interfering signal and produces an output signal that we could subtract, or cancel, from the original sensor signal. Then, if the interfering frequency changes the circuit will still subtract a signal with the correct frequency.

Assuming we can derive a suitable reference signal, there are various methods that allow us to reduce the error between the desired and actual output values. The most difficult decision is what algorithm should be used. Digital filters designed on DSPs are relatively simple to convert to adaptive designs. If the time between samples is sufficiently long, we can compute the new coefficients in parallel with the calculation of the filter output.

The most widely known adaptive algorithm is the Least Mean Square (LMS) algorithm because it is easy to implement and is well understood. There are many different type of LMS algorithms, including normalized LMS, leaky LMS, and sign-error LMS. The implementation of all of these on TMS320Cxx DSPs can be found in Papamichalis (ed.) [1991].

## REFERENCES

There is no shortage of books covering the subject of digital filters. Most of these are advanced and require a good deal of confidence with mathematics. Jackson [1989] may be particularly recommended and was a source of reference for our own writing. The classic digital filter text is probably Hamming [1989]. All DSP textbooks cover the Z-transform, though Jury [1964] is the definitive study. For those looking for an explanation beyond our basic approach, but without the daunting mathematics, try Bateman and Yates [1988], which is a good half-way point.

There are also books that concentrate on adaptive digital filters. Haykin [1991] is a standard text, but complex. Cowan and Grant [1985] is also a comprehensive study. Treichler et al. [1987] offers a practical approach to the subject.

Ahmed, H. and Spreadbury, P.J. [1978]. *Electronics for Engineers*. Cambridge University Press, Cambridge, UK.

Bateman, A. and Yates, W. [1988]. *Digital Signal Processing Design,* Pitman Publishing, London, UK.

Cowan, C.F.N. and Grant, P.M. [1985]. *Adaptive Filters,* Prentice Hall, Englewood Cliffs, NJ.

Hamming, R.W. [1989]. *Digital Filters,* Prentice Hall, Englewood Cliffs, NJ.

Haykin, Simon. [1991]. *Adaptive Filter Theory,* Prentice Hall, Englewood Cliffs, NJ.

Jackson, L.B. [1989]. *Digital Filters and Signal Processing,* Second Edition, Kluwer Academic Publishers, Norwell, MA.

Jury, E.I. [1964]. *Theory and Application of the Z-Transform Method,* John Wiley, New York.

Oppenheim, A.V. and Schafer, R.W. [1975 and 1988]. *Digital Signal Processing,* Prentice Hall, Englewood Cliffs, NJ.

Papamichalis, Panos (ed.) [1991]. *Digital Signal Processing with the TMS320 Family, Volume 3,* Prentice Hall, Englewood Cliffs, NJ.

Parks, T.W. and Burrus, C.S. [1987]. *Digital Filter Design,* Wiley and Sons, New York.

Treichler, J.R.; Johnson, C.R. and Larimore, M.G.[1987]. *Theory and Design of Adaptive Filters,* Wiley and Sons, New York.

Zaks, R. [1981]. *From Chips to Systems,* Sybex, California.

# 5

# Transforming Signals into the Frequency Domain

In previous chapters we have used the description of a signal in both the time domain and the frequency domain interchangeably. For example, in Chapter 4 we used the time domain to demonstrate how a DSP program would be written to implement the filter algorithm, but we always used the frequency domain description to describe the type of filter to be used, e.g., low-pass, band-pass, etc.

We do this generally without thinking. Our intelligent minds have no difficulty thinking in these two directions simultaneously. Unfortunately, the humble DSP is not quite so smart. In order to compute the output of a system for a given input signal, we must provide it with a logical, step-by-step method of computing the result. We are then faced with a dilemma: If the input signal is a sequential series of digital pulses, i.e., a time domain signal, and the system is described by its frequency response, how do we program the DSP?

The answer is quite simple. We either transform the input signal into the frequency domain, or the system response into the time domain. Whichever we choose we shall easily be able to compute our algorithm. Both types of transformations are used in digital signal processing. For example, we often transform the frequency response into the time domain to allow us to build digital filters, as we saw in Chapter 4.

With FIR or IIR filters we produce a time domain representation of the filter response, which we convolve with the input signal to calculate the resulting output. We shall not go into the finer details of transforming the frequency response of a filter into FIR or IIR coefficients since this is covered very expertly by Bateman and Yates [1988].

The other method is to convert the input signal into the frequency domain, which is extremely helpful when we wish to understand the frequency characteristics of a signal. For example, knowing the frequency response of a telecommunications channel is very useful. It allows us to decide what the maximum frequency is that we can transmit and what distortion the signal will receive after passage down the channel.

Another example is in speech analysis. By transforming a speech signal into its frequency components, we can distinguish between speakers and determine the words spoken. This is very useful in speech recognition and identification, two applications that are gaining prominence as the performance of DSPs increases. Chapter 6 covers several vocoding algorithms that use frequency transforms to allow the system to comprehend the information content in speech and reproduce it.

The other most obvious example of transforming a time domain signal into the frequency domain is in spectrum analyzers, which are now in general use in most electronic laboratories. Spectrum analyzers may be used to examine the output from sensors attached to mechanical structures, e.g., bridges, where a significant change in the frequency response can mean excess stress on some part of the structure and breakdown in the near future. If you wish to read more in depth on the design of spectrum analyzers, Martin [1991] offers a good insight.

In this chapter, we shall examine how signals are transformed from one domain to the other. Although previously we have avoided using equations to explain any DSP techniques, in this chapter we shall study how we derive the relevant transforms. Most other DSP textbooks assume the reader already understands the derivation of the transforms, which can immediately alienate the casual explorer. We hope that after understanding the basics you will be able to move on to the more advanced texts with a newfound confidence.

We shall then look at why the advent of sophisticated digital signal processors has made these relatively old transforms usable. As we said earlier, Jean Baptiste Joseph Fourier derived his classic formulas back in 1822! We shall look at some of the newer implementations of Fourier's formulas and what features you should look for in a DSP if you intend to use it for any of the purposes outlined above.

The final part of this chapter will examine another type of frequency transform that is becoming very popular in image processing applications.

Unfortunately, it is not practical to explain frequency transforms without a fair bit of mathematics. Although there are a lot of equations in this chapter we hope that we can provide a logical explanation of the subject and not lose too many readers along the way. If you don't want to understand the math, but just need to implement a discrete Fourier transform (DFT), fast Fourier transform (FFT) or discrete cosine transform (DCT), we suggest you refer directly to Papamichalis (ed.) [1991].

## THE PHASOR MODEL

As a starting point, we need a simple method of describing a signal. We shall use the phasor model. Figure 5.1 shows a single phasor.

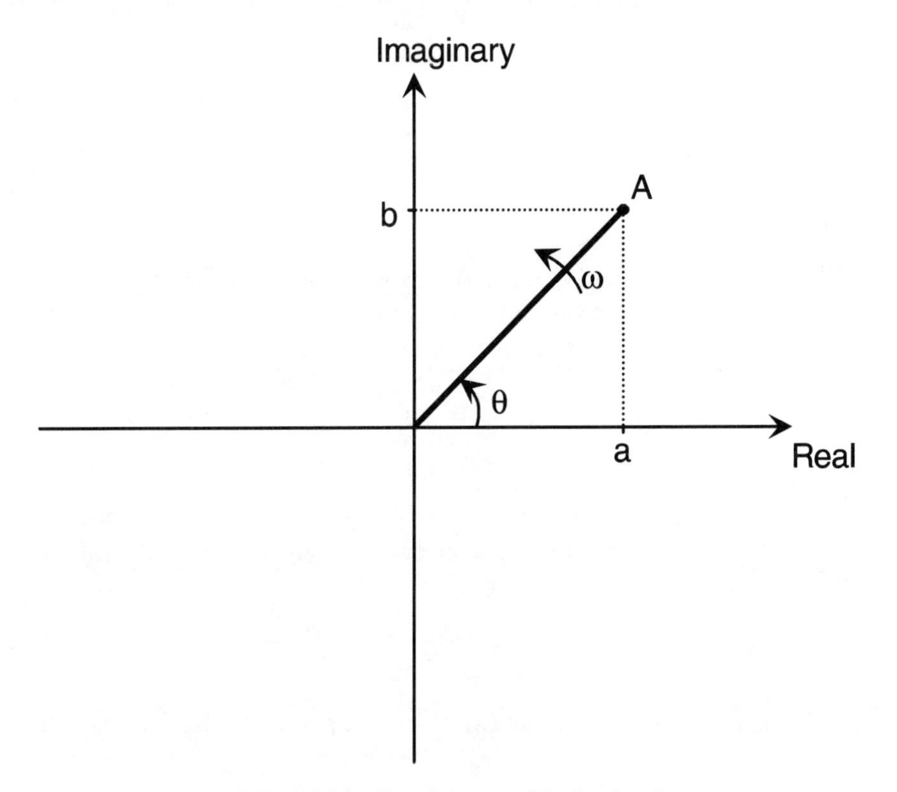


**FIGURE 5.1.** The phasor model of a signal

A phasor is actually a vector rotating in the complex plane, with a magnitude of A and a rotational speed of $\omega$ radians/sec. If we take a snapshot in time we can see that the signal at that time, $x(t)$, is given by:

$$\begin{aligned} x(t) &= (\text{real co-ordinate}) + j(\text{imaginary co-ordinate}) \\ &= a + jb \end{aligned}$$

where:

$$A = \sqrt{a^2 + b^2}$$

and:

$$\theta = \omega t = \tan^{-1} \frac{b}{a}$$

This is one method of writing a complex number (remember $j^2 = -1$), called the rectangular form. The other method is the polar form where:

$$x(t) = Ae^{j(\omega t)}$$

and:

$$e^{j(\omega t)} = \cos(\omega t) + j\sin(\omega t)$$

These basic equations give rise to many of the trigonometric functions that most of us had to endure at school. Remember also that $\omega$ is related to frequency by the following equation:

$$\omega = 2\pi f$$

Finally $\pi$ radians is equivalent to 180°.

Our phasor description can easily be extended to discrete time, or digital systems, where the signal occurs only at specific intervals in time defined by the sampling interval $T_s$:

$$x(n) = Ae^{j(n\omega T_s)}$$

So, instead of the continuous variable time $t$, we now have a discrete variable $n$, so that the phasor advances in jumps of $T_s$.

Taking either the continuous or discrete case, if we have an initial value for $x$ of:

$$x(0) = Ae^{j(\alpha)}$$

we can derive the general form of both equations as follows:

$$x(t) = Ae^{j(\omega t + \alpha)}$$

$$\text{or } x(n) = Ae^{j(n\omega T_s + \alpha)}$$

These simple equations form the basis of all the following analysis. We shall now look at how we use this model to describe signals that we are familiar with: sine and cosine waves.

## MODELING SINUSOIDS

Going back to the description of $e^{j(\omega t)}$, we can rewrite it as:

$$e^{j\theta} = \cos\theta + j\sin\theta$$

also:

$$e^{-j\theta} = \cos\theta - j\sin\theta$$

where:

$$\theta = (\omega t + \alpha) \text{ or } (n\omega T_s + \alpha)$$

From these two equations we can derive the following relationships:

$$\cos\theta = \frac{e^{j\theta} + e^{-j\theta}}{2}$$

$$\sin\theta = \frac{e^{j\theta} - e^{-j\theta}}{2j}$$

This means that a general sine or cosine signal, $x(t)$, can be described by the sum of two phasors. For example:

$$x(t) = R\cos(\omega t + \alpha)$$

$$= \frac{R}{2}(e^{j(\omega t + \alpha)} + e^{-j(\omega t + \alpha)})$$

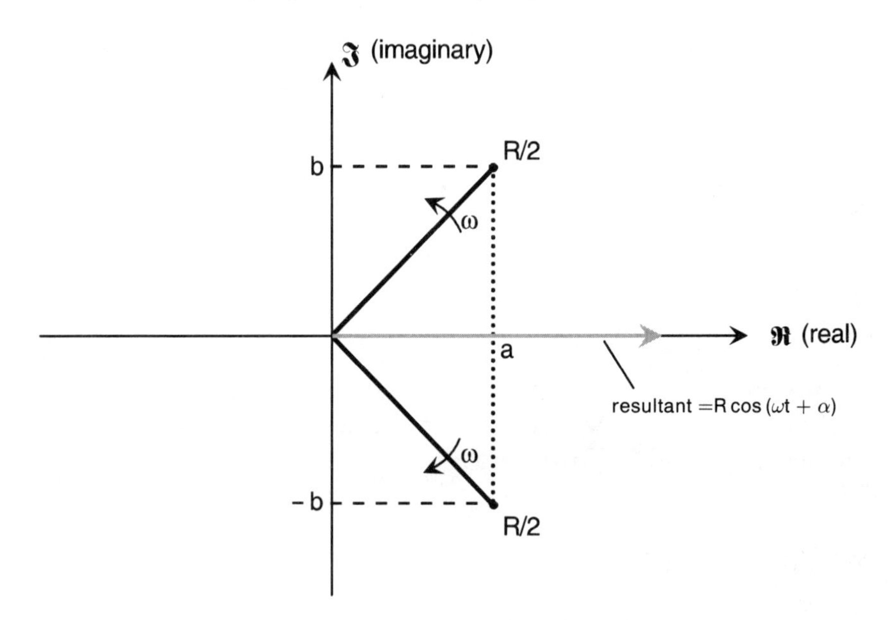


**FIGURE 5.2.** Phasor representation of a cosine signal

Therefore, our cosine signal can be represented by two phasors that form a conjugate pair. This means they have the same real value ($a$) and equal and opposite values of $b$; see Figure 5.2. We could work out the phasor form of a sine wave and we would find it also consists of a conjugate pair of phasors, but the sign would be different.

Remembering that our phasors are rotating in either the positive (clockwise) or negative (counterclockwise) directions, we can derive the very interesting property that all real signals must be made up of conjugate pairs of phasors so that the sum of the vectors always lies on the real axis.

## FOURIER SERIES

The discussion so far has related to simple sine or cosine waves of a single frequency, but more intricate waveforms can also be split into many cosine or sine waves. For example, a rectangular pulse train consists of an infinite number of sine waves of varying amplitude. Therefore, we can describe any complex periodic signal as the sum of many phasors. One method of describing a signal in this way is called the Fourier series, which assumes that the set of phasors have frequencies that are multiples of some fundamental frequency, $f_0$ (or angular frequency, $\omega_0$):

$$x(t) = \sum_{k=-N}^{N} C_k e^{j(k\omega_0 t)}$$

Any periodic signal can be represented by a Fourier series provided that $N$ is big enough. The individual frequency components are known as harmonics.

We can make the Fourier model more general by using phasors whose frequencies are not harmonically related, which is generally the case when the complex signal is not periodic (most cases in real applications):

$$x(t) = \sum_{k=-N}^{N} C_k e^{j(\omega_k t)}$$

Any arbitrary waveform may be represented by a Fourier series of this general type.

## DISCRETE FOURIER SERIES

We now need to translate these continuous time equations into the discrete or digital domain to enable us to derive some useful formulas for our DSPs. The above analysis can be extended to discrete time systems. All that is necessary is to replace the continuous function, $t$, with one that progresses in jumps of $\omega_0 T_s$, so for the periodic case:

$$x(n) = \sum_{k=-N}^{N} C_k e^{j(k\omega_0 T_s n)}$$

It is interesting to note that when the phase jump for the $k$th harmonic is given by:

$$k\omega_0 T_s = 2\pi m$$

where $m$ is an integer, the phase is indistinguishable from when $k = 0$. This happens because $2\pi = 360°$ and it occurs when:

$$k\frac{\omega_0}{\omega_s} = 1, 2, 3, 4 \ldots.$$

Taking a little time to think about this, we can see it means the frequency response of a discrete signal is periodic with a period of $\frac{1}{T_s}$. This is the fact we assumed back in Chapter 3 when describing the process of sampling to produce a discrete signal and is a very important feature in digital signal processing.

We have now used our simple phasor model to describe a general discrete signal. Using this description we can go on to explain how we can transform between the time and frequency domains.

## NONPERIODIC SIGNALS – THE FOURIER TRANSFORM

In real applications most signals are not periodic, so we must modify our Fourier series to comprehend this. Let's consider the general Fourier series where all the frequencies are harmonically related, i.e.:

$$\omega_k = k\omega_0$$

The fact that the final signal is not periodic may be represented by:

$$\omega_0 \rightarrow 0$$

This equation simply states that there is no "least common denominator" in the frequencies of all our separate phasors. Then the number of phasors tends towards infinity and our summation becomes an integral:

$$x(t) = \frac{1}{2\pi} \int_{-\infty}^{+\infty} X(\omega) \cdot e^{j(\omega t)} \cdot d\omega$$

In the above equation we have assumed that the signal amplitude can be defined as a function of frequency ($\omega$), i.e., $X(\omega)$. The "reverse" equation defining $X(\omega)$ is given by:

$$X(\omega) = \int_{-\infty}^{+\infty} x(t) \cdot e^{j(-\omega t)} \cdot dt$$

Therefore, we now have an equation that allows us to calculate the amplitude response of a continuous signal in the frequency domain using its time domain response. These two equations are called the

Fourier transform pair. They are very useful formulas for mathematicians but unfortunately it isn't possible to implement them on a DSP. To do that we must produce the discrete form, which we shall cover shortly.

Obviously, an infinite number of measurements are now necessary to determine the function $X(\omega)$. In practice, the Fourier transforms are not actually calculated; we simply use tables of Fourier transform pairs printed in most mathematics or DSP textbooks (see Lynn and Fuerst [1989]).

## THE DISCRETE FOURIER TRANSFORM (DFT)

In order to find the discrete equivalent of the Fourier transform we must replace the continuous variable $t$ by the discrete variable $nT_s$. Outside $\pm\pi/T_s$ the spectrum repeats itself; therefore we could change the limits to this value. For convenience though, we will make the variable of integration $\omega T_s$ and the integral now becomes:

$$x(n) = \frac{1}{2\pi} \int_{-\pi}^{+\pi} X(\omega) \cdot e^{j(\omega T_s n)} \cdot d(\omega T_s)$$

The reverse transform is:

$$X(\omega) = \sum_{n=-\infty}^{\infty} x(n) \cdot e^{j(-\omega T_s n)}$$

Notice that the second equation still uses a discrete summation rather than an integral. We would expect this because we know that $x(n)$ is only valid at $nT_s$ instants in time. These two equations form the DFT pair, which we shall look at in a little more detail. What we now have is a pair of equations that allow us to transform discrete, or digital, signals between the time and frequency domains.

The spectrum produced using the DFT has some rather interesting properties. For example, it is periodic with frequency $\omega_s$ – as we explained before. Also, for a real signal, because the phasors must occur in conjugate pairs, the spectrum must always have even symmetry along the real axis and odd symmetry on the imaginary axis. This simply means that if we know we are dealing with a real signal, the amount of information we need to remember about the frequency spectrum is less, since it is repeated (see Oppenheim and Schafer [1975 & 1988]).

Using these two equations we can now determine the coefficients of an FIR or IIR filter. Alternatively, we can calculate the frequency response of a digitized speech signal, or any of the other applications we looked at back at the start of this chapter. As with the Fourier transform pair, we can also look up the DFTs of $x(t)$ and $X(\omega)$ (see again Lynn and Fuerst [1989]), but we shall progress a little further with this discrete form of the equations and see how we use our DSPs to perform this function.

## PRACTICAL CONSIDERATIONS

Taking stock of what we have covered so far, we have introduced a method of describing a signal that varies with time, called the phasor model. We related this model to sine and cosine waves and introduced the Fourier series, which tells us that any periodic signal can be represented by a number of harmonically related sine waves. After making our Fourier series more general by removing the relationship between the harmonics we then applied our discrete form of the equation, which has finally given us the DFT pair.

So now we have a pair of equations that will allow us to transform between time and frequency domains for any discrete signal, but how do we program our DSP to perform a DFT? What problems can we expect when we try to design a DFT processing algorithm?

Referring back to the DFT equation for $X(\omega)$, there are two obvious problems: First, in the real world an infinite summation is not possible, since we would never produce an answer. Second, we are limited in the time we have to compute an output – even with a DSP. Therefore, we are restricted in the number of frequencies at which we can perform the mathematics.

The first problem is easy to overcome: We must take only a section of the input values $x(n)$. This is usually referred to as windowing and it is also used in many other applications, e.g., vocoders (see Chapter 6). If we then perform our DFT on the windowed $x(n)$, the resulting spectrum is given by:

$$X_N(\omega) = X(\omega) * W(\omega)$$

where $X_N(\omega)$ denotes the windowed spectrum with $N$ as the number of samples used, $W(\omega)$ denotes the spectrum of the window and the $*$

denotes the convolution of $W(\omega)$ and $X(\omega)$. Convolution is a fancy name for a simple operation, as we saw back in Chapter 4. To refresh our memories we can rewrite the equation:

$$X_N(\omega) = \sum_{r=0}^{N-1} X(r) \cdot W(N-r)$$

As shown in the next chapter, the ideal window would have a rectangular spectrum in the frequency domain. Remembering that we said the frequency spectrum will be periodic with frequency $\omega_s$, we can see that a rectangular spectrum in the frequency domain implies that we shall not cause any interference between the adjacent "lobes" of the response. Unfortunately, a rectangular frequency response is practically impossible to build. We must therefore always settle for some compromise (see Chapter 6).

Let's assume we have chosen a windowing function. What we need to know now is the effect of using only a limited number of frequencies and how many frequencies are needed to maintain sufficient accuracy. Basically there is no simple answer. In general, the optimum number of phasors is equal to the number of initial points in $x(n)$, i.e., $N$. The simplest way to imagine this is to assume that our windowed portion of $x(n)$ consists of one period of a long sequence (see Figure 5.3), with period $NT_s$ and frequency $\omega_s/N$. If we do this we can treat the sequence as a Fourier series and we can see that the spectrum will consist of $N$ phasors.

If you move on to the more detailed books on DSP you will probably not see the DFT pair written as shown earlier. They generally use another "variable" in the equation called a twiddle

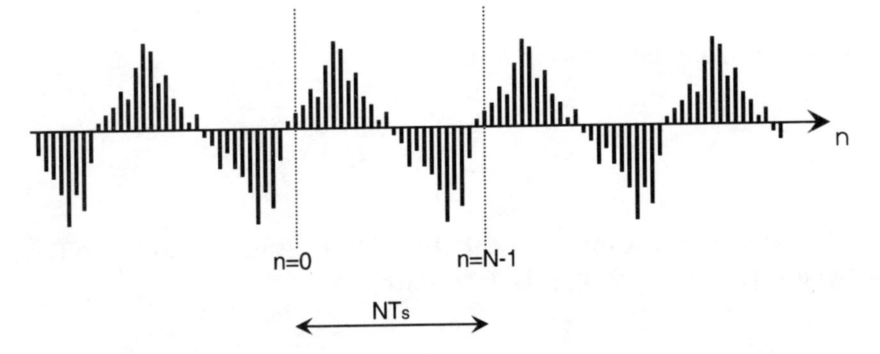


**FIGURE 5.3.** Windowed portion of signal as one period of a sequence

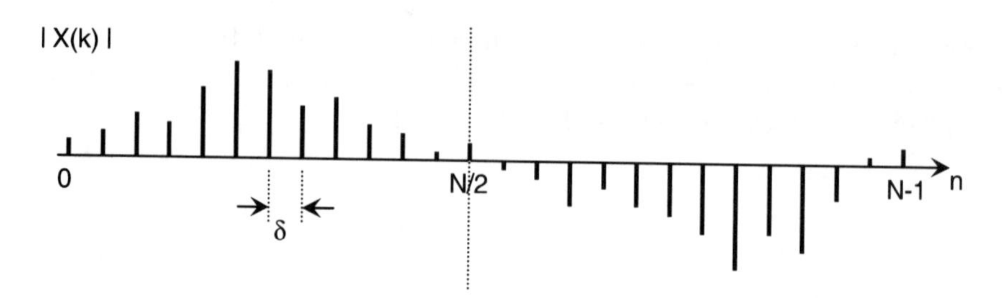


**FIGURE 5.4.** Phasors spaced by $\delta$

factor. Let's go back a little and see where this new function comes from. We noted earlier that the spectrum of a DFT is repetitive with period $\omega_s$, so if we say that the phasors are spaced by $\delta$ (Figure 5.4), then:

$$N\delta = \omega_s$$

This has allowed us to digitize our frequency scale so that our spectrum can now be written in terms of $k$ instead of $\omega$:

$$X(k\delta) = K(\delta) = \sum_{n=0}^{N-1} x(n) \cdot e^{j(-k\delta T_s n)}$$

Remembering that:

$$\omega_s = \frac{2\pi}{T_s}$$

and that:

$$\delta = \frac{\omega_s}{N}$$

we can rewrite the equation to be:

$$X_N(k) = \sum_{n=0}^{N-1} x(n) \cdot e^{-j\left(\frac{2\pi kn}{N}\right)}$$

This is the form of the practical DFT that is generally used and the twiddle factor, called $W_N$, is defined as:

$$W_N = e^{-j\left(\frac{2\pi}{N}\right)}$$

So finally we can write the DFT pair in its most common form:

$$X_N(k) = \sum_{n=0}^{N-1} x(n) \cdot W_N^{kn}$$

and:

$$x(n) = \frac{1}{N} \sum_{k=0}^{N-1} X_N(k) \cdot W_N^{-kn}$$

## FAST FOURIER TRANSFORM (FFT)

The DFT is a very efficient method of determining the frequency spectrum of any signal. The only drawback with the technique is the amount of time required to compute the output. This is because both indices, $k$ and $n$, must progress through $N$ values to produce the full range of output phasors and therefore $N^2$ computations must be performed.

Looking at the previous two equations we can see that each part of the equation will be a multiply-and-add function, so a DFT with $N = 1000$ (1000 point DFT), will use $10^6$ machine cycles even on a DSP. Using a DSP with a cycle time of 50ns this would take 0.05s and then the highest sample rate we could accept would be 20Hz!

If we reexamine the twiddle factor $W_N$:

$$W_N = e^{-j\left(\frac{2\pi}{N}\right)}$$

we can see that the same values of $W_N$ are calculated many times during the DFT since $W_N$ is a periodic function with a limited number of distinct values. The aim of the fast Fourier transform (FFT) and its inverse, the IFFT, is to use this redundancy to reduce the number of calculations.

The generic term *fast Fourier transform* covers many different algorithms with different features, advantages and disadvantages. For example, an FFT designed to be optimum when used with a high-level language will probably not work very well on a fixed-point DSP. Nevertheless, all FFTs use the same approach to reducing the long algorithm into a number of successively shorter and simpler DFTs. We shall look at one method of reducing the DFT to a more sensible FFT in a little detail here. Although this requires a fair amount of mathematics it will allow us to pick out the reasons why many DSPs are designed to implement FFTs as efficiently as possible.

The first step in this reduction is to split the input signal $x(n)$ into several shorter interleaved sequences. This is usually referred to as decimation-in-time. If we take our original signal with $N$ values, let us split it first into two sequences, one of the odd numbers and one of the even numbers:

$$X_N(k) = \sum_{n=0}^{N-1} x(n) \cdot W_N^{kn}$$

becomes:

$$X_N(k) = \sum_{r=0}^{\frac{N}{2}-1} x(2r) \cdot W_N^{2rk} + \sum_{r=0}^{\frac{N}{2}-1} x(2r+1) \cdot W_N^{(2r+1)k}$$

or:

$$X_N(k) = \sum_{r=0}^{\frac{N}{2}-1} x(2r) \cdot (W_N^2)^{rk} + W_N^k \sum_{r=0}^{\frac{N}{2}-1} x(2r+1) \cdot (W_N^2)^{rk}$$

Going back to our definition of $W_N$:

$$W_N = e^{j(-\frac{2\pi}{N})}$$

so:

$$W_N^2 = e^{j(-\frac{2\pi}{N}2)} = e^{j(-\frac{2\pi}{N/2})} = W_{N/2}$$

and our equation becomes:

$$X_N(k) = \sum_{r=0}^{\frac{N}{2}-1} x(2r) \cdot W_{N/2}^{rk} + W_N^k \sum_{r=0}^{\frac{N}{2}-1} x(2r+1) \cdot W_{N/2}^{rk}$$

which can be written:

$$X_N(k) = G(k) + W_N^k H(k)$$

where $G(k)$ is the DFT of the even numbered points and $H(k)$ is the

DFT of the odd numbered points. Unfortunately, we must multiply the odd DFT by $W_N^k$ before we can add it to $G(K)$. We have now expressed the original DFT in the form of two smaller DFTs of length $\frac{N}{2}$. Where $N = 1000$, the calculation of these will take $500^2 + 500^2 + 500 = 50{,}500$ rather than the $10^6$ of the original DFT.

Assuming that the transform length is an integer power of 2, we can take this decimation further and break these two sequences themselves into two more, the limit to this process being very simple 2-point DFTs where we arrive at one of the most commonly used forms of FFT, the radix-2. As an example, let us take a short sequence where $N = 8$, i.e., $x(n)$ is defined for:

$$n = \{0, 1, 2, 3, 4, 5, 6, 7\}$$

Decimation once gives two sequences:

$$n = \{0, 2, 4, 6\} \text{ and } \{1, 3, 5, 7\}$$

Decimation again gives four sequences:

$$n = \{0, 4\}\,\{2, 6\}\,\{1, 5\} \text{ and } \{3, 7\}$$

Now our FFT is calculated by initially performing four 2-point DFTs on the above pairs of sample values. Although this simplifies the mathematics tremendously, we still must remember the twiddle factors which arise every time we decimate the sequence. These factors are also different at each stage of decimation. The aim of a good FFT algorithm is to be able to incorporate these extra factors into the mathematics without encumbering the DSP too much.

Let's take an even simpler example where $N = 4$ and look at the math involved:

$$X_4(k) = \sum_{n=0}^{3} x(n) \cdot W_4^{kn}$$

$$= \sum_{r=0}^{1} x(2r) \cdot W_2^{rk} + W_4^k \sum_{r=0}^{1} x(2r+1) \cdot W_2^{rk}$$

$$= [x(0) + x(2) \cdot W_2^k] + W_4^k [x(1) + x(3) \cdot W_2^k]$$

But we can rewrite the $W_N$ so that:

$$W_2^k = e^{j(-\frac{2\pi}{2}k)} = e^{j(-\frac{2\pi}{4}2k)} = W_4^{2k}$$

Then our FFT becomes:

$$X_4(k) = [x(0) + x(2) \cdot W_4^{2k}] + W_4^k[x(1) + x(3) \cdot W_4^{2k}]$$

So writing out all the values for $k$ longhand, we get:

$$X_4(0) = [x(0) + x(2) \cdot W_4^0] + W_4^0[x(1) + x(3) \cdot W_4^0]$$
$$X_4(1) = [x(0) + x(2) \cdot W_4^2] + W_4^1[x(1) + x(3) \cdot W_4^2]$$
$$X_4(2) = [x(0) + x(2) \cdot W_4^0] + W_4^2[x(1) + x(3) \cdot W_4^0]$$
$$X_4(3) = [x(0) + x(2) \cdot W_4^2] + W_4^3[x(1) + x(3) \cdot W_4^2]$$

Notice that:

$$W_4^4 = e^{j(-\frac{2\pi}{4})4} = 1 = W_4^0$$

and:

$$W_4^6 = e^{j(-\frac{2\pi}{4})6} = -1 = W_4^2$$

It may be easier to visualize this 4-point DFT using a signal flow graph as shown in Figure 5.5 where the numbers in the circles represent the power of $W_4$ that is required in that stage. If there is no symbol, then no multiplication is necessary. The signal flow graph illustrates one of the main features of any type of FFT, the butterfly. Each section is made up of several butterflies, as shown in Figure 5.6.

The generally accepted form of the butterfly is when the multiplication factors are $x = 0$ and $y = -1$. This helps when looking at many books that simply make this assumption and draw the butterfly as shown in Figure 5.7. In the case where we have extra twiddle factors we can draw the butterfly as shown in Figure 5.8. The butterfly now only consists of simple additions and subtractions, since the twiddle factors are all either $+1$ or $-1$. When we have managed to reduce the equations to this simple format it becomes extremely easy to design a program to perform the FFT.

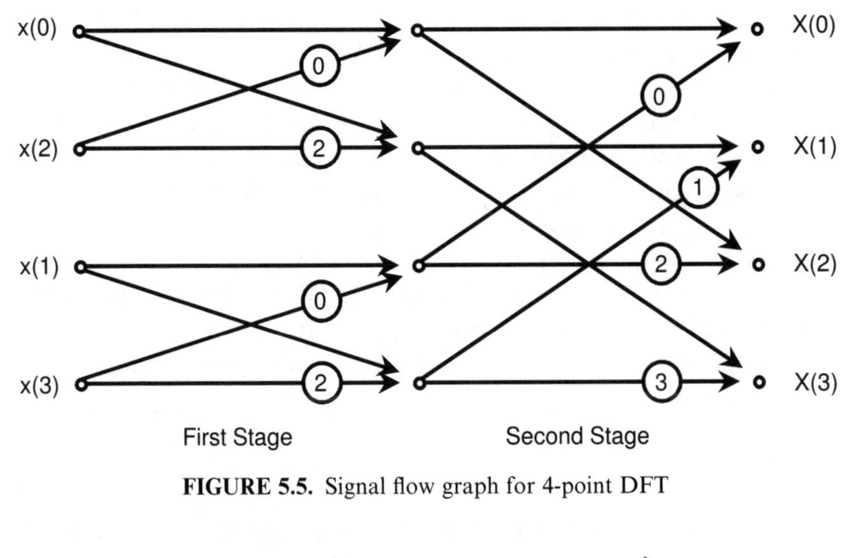


**FIGURE 5.5.** Signal flow graph for 4-point DFT

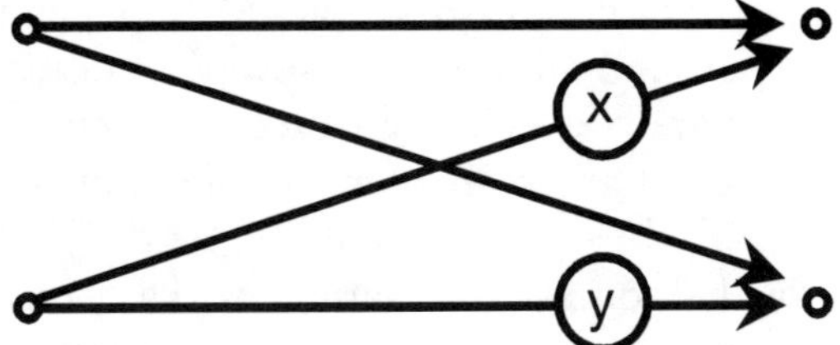


**FIGURE 5.6.** The basic FFT butterfly

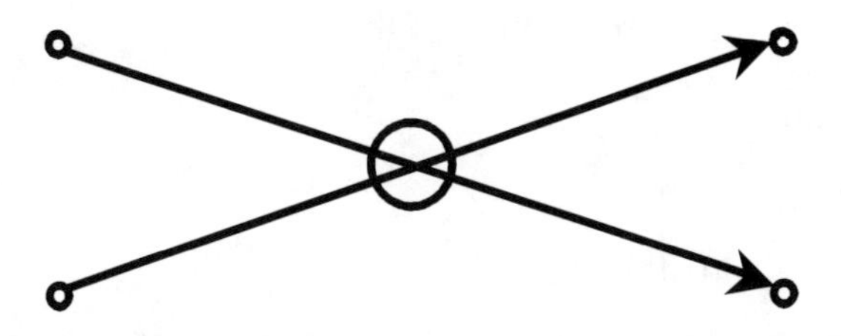

**FIGURE 5.7.** Alternative symbol for butterfly

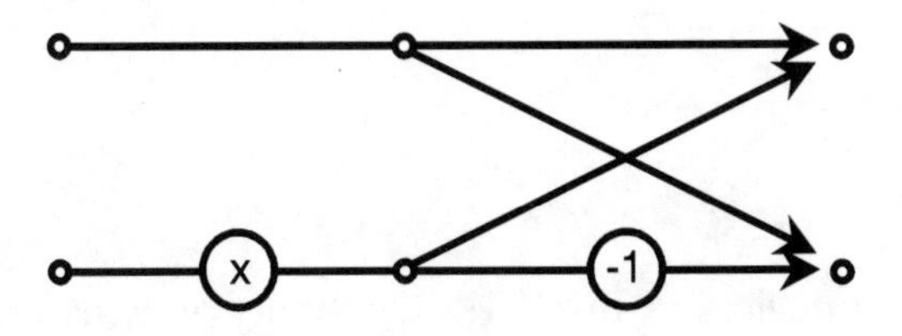


**FIGURE 5.8.** General case of butterfly with extra twiddle factors

A single-chip DSP is ideal for the purpose of performing FFTs since it can perform the mathematics quickly. DSP devices also have another important feature, which we shall cover now. With any FFT, in order to get the output values in ascending order, we must arrange the input sequence in a shuffled manner. It can be shown that the consequence of not reordering the input sequence would be a shuffled output sequence.

Looking at how an FFT can be implemented on a DSP, we must have two data tables, one for the input values and the second for the output values. With our 4-point FFT, to obtain the output in ascending order the input values must be loaded in the table in the following order:

$$n = \{0, 2, 1, 3\}$$

This is trivial for a 4-point FFT, but if we move on to 512-point, 1024-point or larger FFTs it isn't so simple. Most general-purpose DSPs now offer a way around this called bit-reversed addressing. Using this, the initial values are conveniently stored in ascending order and an addressing mode is chosen to pick out the values in the required shuffled order.

This is how the address calculation works: Consider a three-bit address, which represents $2^3$, or 8 possible memory locations. We are therefore looking at an 8-point FFT. After starting at zero, what we are going to do is to add half the FFT length at each address access, but do so carrying from left to right:

$$\begin{aligned}
&\text{start at } 000 && = x(0) \\
&000 + 100 && = 100 = x(4) \\
&100 + 100 && = 010 = x(2) \\
&010 + 100 && = 110 = x(6) \\
&110 + 100 && = 001 = x(1) \\
&001 + 100 && = 101 = x(5) \\
&101 + 100 && = 011 = x(3) \\
&011 + 100 && = 111 = x(7)
\end{aligned}$$

In the newer devices in TI's TMS320 family of general-purpose DSPs this address manipulation is done as a background task. We can therefore benefit from the simple table structure without any time penalty for using bit-reversed addressing. This is possible with the devices that have arithmetic units associated with the address generators. It is these special address ALUs that perform the sequence shown above. Using these modern DSPs it has become possible to implement a 1024-point FFT in well under 5ms. Because this calculation time has been reduced to such a small value, we can now perform real-time tasks with DSPs. There are also specific hardware processors available whose sole function is to perform FFTs. Obviously these devices are capable of even faster calculation.

Going back to our 4-point FFT example, let's look at the values of $W$ we have in the butterflies:

$$\begin{aligned} W_4^1 &= e^{j(-\frac{2\pi}{4})} \\ &= e^{-j90^\circ} \\ &= \cos(-90^\circ) + j\sin(-90^\circ) \\ &= -j \end{aligned}$$

and:

$$W_4^2 = (W_4^1)^2 = j^2 = -1$$

Finally:

$$W_4^3 = W_4^1 W_4^2 = j$$

Therefore, we can simplify our signal flow graph as shown in Figure 5.9. As we noted earlier, multiplication by $-1$ is simply a subtraction and by $\pm j$ is simply a conversion between real and imaginary numbers. Therefore, the DFT is very simple and it is possible to convert the diagram to the standard form as shown in Figure 5.10. The mathematics have now been reduced as far as possible, to only one twiddle factor of $-j$.

This is just one way of describing the FFT. Unfortunately, in other books you will find many different methods of saying the same thing. Hopefully, now that we have described the functions and the nature of the twiddle factor $W$, you will have no problems interpreting these assorted methods.

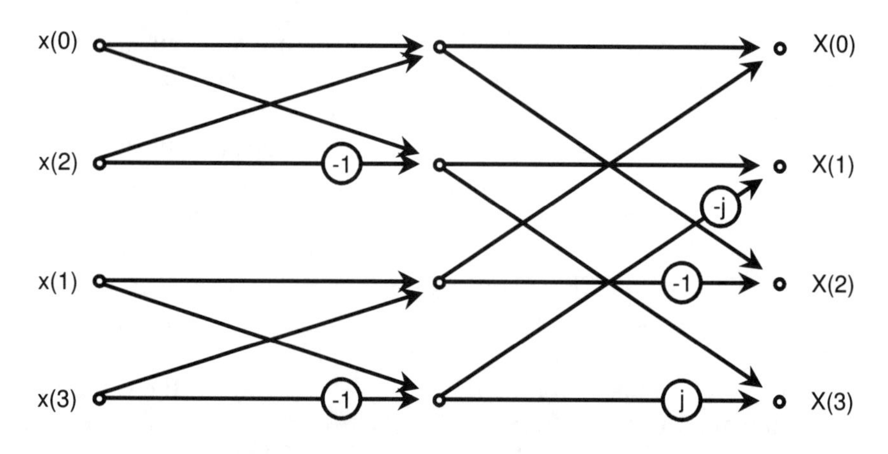


**FIGURE 5.9.** Simplified flow graph for 4-point DFT

As implied earlier, the advantage of using decomposition into smaller DFTs is the reduction in the number of operations required to perform the final function. If we decompose an $N$-point DFT right down to 2-point DFTs, the number of processing operations is reduced from $N^2$ to $N \log_2 N$. Taking an example of a 512-point DFT to illustrate this point, we can see we have reduced the number of operations from 260k to 4.6k, an improvement of over 50 times.

Remember that the term *fast Fourier transform* is a general one used for many methods of reducing the amount of computation required to perform DFTs. Here we have looked only at the radix-2 FFTs, which

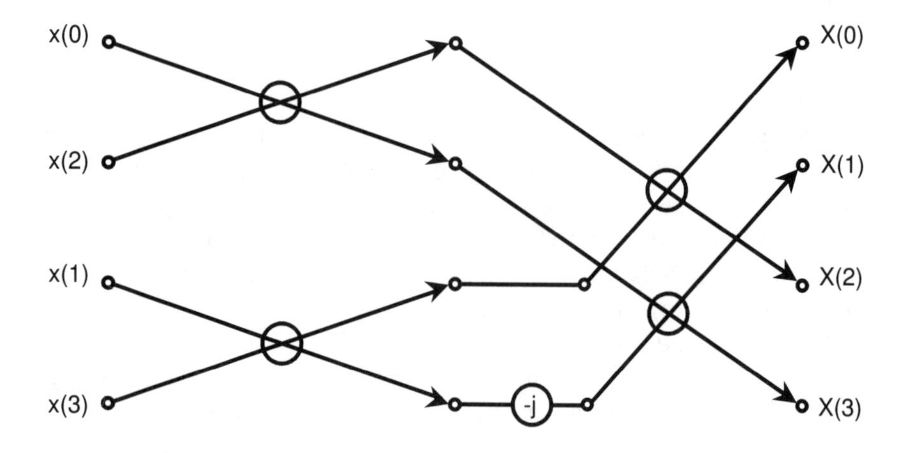


**FIGURE 5.10.** Standard form of flow graph for 4-point DFT

are the most popular but not the only method. It is simple to see from the above example how the decomposition to 2-point DFTs can easily be achieved with any size of FFT that is an integer power of 2. Obviously, if that is not the case we should have to use a different form.

We shall mention one special case here. If you want to read more on DFTs and FFTs, the book by Burrus and Parks [1985] covers a lot of the methods and introduces some programs for the TMS320 family of DSPs. Additional programs that implement FFTs on the TMS320 DSPs can be found in the range of TI applications manuals.

## THE GOERTZEL ALGORITHM

If we are only interested in the spectrum of a signal at one particular frequency point, we can use a special form of the DFT called the Goertzel algorithm. One example of the use of this algorithm is in a DTMF (Dual Tone Multi Frequency) encoder and decoder. The main advantage of this algorithm is that the coefficients in the equation are fixed for a particular frequency point. This makes the calculations much simpler. A full description of this algorithm and coding for the TMS320C17 can be found in the family of TI applications manuals.

## DISCRETE COSINE TRANSFORM (DCT)

In this section we shall briefly look at one other type of frequency transform called the Discrete Cosine Transform (DCT). This transform has also gained a lot of interest in the last few years as it has become possible to implement sizeable functions real time with the latest DSPs.

The DCT is just one of a number of suboptimal image-coding techniques. The Karhunen-Loeve transform is optimal in the mean square error sense. In real systems this is rarely used because it is very computationally intensive. DCTs are becoming a commonly used substitute, as we shall see in Chapter 6.

A DCT can provide compression ratios in the order of 20 : 1, which means reducing a $1024 \times 1024$ pixel, 8-bit image to a 50k byte file. This is because the DCT renders the image into frequency components using a single pipelined operation, in comparison to, for example, sub-band coding, where a block of parallel band-pass filters are required.

We invariably break up a large image into $8 \times 8$ pixel blocks before we perform DCTs on an image. One reason is that we wish to use the DCT to exploit the redundancy in a set of pixels. Any one pixel in a picture is likely to be closely related to the four pixels that surround it and similarly each of those four pixels are likely to bear some relation to their four nearest neighbors. But the original pixel is unlikely to be related to one a long distance away. Therefore, by splitting up the image we hope to form groups of pixels that are statistically related with a consequently high level of redundancy.

The second reason for breaking the picture into smaller blocks is to reduce the amount of calculation required. The DCT is similar to the DFT and is nontrivial, as we shall see shortly. It is impractical to try and perform the DCT on the whole image – the calculation would take too long for real-time image processing.

After performing a DCT on a block of pixels we have a set of coefficients that are frequency dependent. If we take a television picture or photograph, we generally find that the low-frequency components (including dc) are quite large in comparison with the high-frequency signals. We can often either ignore the higher-frequency coefficients or use some form of variable-length coding scheme. Variable-length coding means that more bits are assigned to the coefficients that contain the most crucial information, i.e., the low-frequency components, while the rest use a reduced number of bits. In this way we can reduce the number of bits required to transmit the picture and so we have reduced the signal bandwidth.

Reduction in bandwidth is the ultimate aim of most image compression techniques. To go back to something we mentioned earlier, the $1024 \times 1024$, 8-bit image requires around 50k bytes after a DCT has been performed. For a picture rate of 25 frames per second, which is prevalent in many videophones, we would need a channel capable of transmitting the resultant signal at a rate of 1.25Mbyte/s, assuming we didn't need any time to decode the image at the other end!

If we split the image into smaller blocks, the amount of information after the DCT was performed would be smaller, so the time taken to perform the inverse DCT (IDCT) would be less. Also, with the increased correlation between pixels, we would be able to reduce the number of coefficients that needed to be transmitted. The total data rate would then be lower.

Let's take a look at the mathematics. The DCT of the discrete series $x(n) = 0, 1, 2, 3, 4 \ldots (N-1)$, is represented by the following equation:

$$X_c(k) = \sqrt{\frac{2}{N}}\, C(k) \sum_{n=0}^{N-1} x(n) \cdot \cos \frac{(2n+1)k\pi}{2N}$$

for $k = 0, 1, 2, 3, 4, \ldots (N-1)$. Where:

$$C(k) = \frac{1}{\sqrt{2}} \quad \text{for } k = 0$$

$$\text{and } C(k) = 1 \qquad \text{for } k = 1, 2, 3, \ldots, N-1$$

Following on, the IDCT is given by:

$$x(n) = \sqrt{\frac{2}{N}} \sum_{k=0}^{N-1} C(k) \cdot X_c(k) \cdot \cos\left[\frac{(2n+1)k\pi}{2N}\right]$$

Obviously, these are all one-dimensional (1-D) transforms for simplicity. References to 2-D transforms can be found in Stafford [1980].

The DCT is related to the DFT, as we might expect. This can be proven using a double-length DFT as shown in the article by Chitprasert and Rao [1990]. This relationship is sometimes used to allow the designer to use an FFT to perform a DCT on a sequence where $N$ is a power of 2. On large DCTs this may be an advantage, but we have already stated that for most purposes it is acceptable to split the image into smaller blocks.

In any application where we cannot tolerate any errors introduced by splitting up the picture, we may need to use the DFT method. An example is in medical imaging systems where the received signal is often very small so signal-to-noise ratios are poor and certain features may be very difficult to distinguish.

In recent years more efficient fast cosine transforms (FCTs) have begun to be developed, but these are at a relatively early stage. Nevertheless, we can see by the similarities of the DCT to the DFT how they will be developed. An example of the implementation of a DCT on a TMS320C30 DSP can be found in Papamichalis (ed.) [1991].

In videophones, the effects of splitting the picture are not too noticeable by the human eye and DCTs are now widely used.

However, they are rarely referred to directly since they form only a part of the overall video compression algorithm. Image compression is described in more detail in Chapter 6.

## REFERENCES

There are many books on the subject of Fourier transforms in all their guises. Most tend toward the mathematical, since the abstract concepts defy all attempts to rationalize their existence. All standard DSP textbooks include chapters on the subject, e.g., Oppenheim and Schafer, Lynn and Fuerst, etc. Alternatively, there are books devoted solely to the subject such as like Burrus and Parks. Tomes on DCTs are more difficult to find. Although all our standard DSP books do cover them lightly, they do not offer any insight into how they work. Pratt [1978] and Jayant and Noll [1984] are a little better than most; otherwise stick to application notes on how to implement the formulas.

Bateman, A. and Yates, W. [1988]. *Digital Signal Processing Design*, Pitman Publishing, London, UK.

Burrus, C.S. and Parks, T.W. [1985]. *DFT/FFT and Convolution Algorithms,* Wiley Interscience, NY.

Chan, K.K. [1988]. "Implementation of Fast Cosine Transforms with Digital Signal Processors for Image Compression," *SPIE Vol 914 Medical Imaging II.*

Chitprasert, B. and Rao, K.R. [1990]. "Discrete Cosine Filtering," *Proceedings of the 1990 IEEE International Conference on Acoustics, Speech and Signal Processing.*

Jayant, N.S. and Noll, P. [1984]. *Digital Coding of Waveforms*, Prentice Hall, Englewood Cliffs, NJ.

Lynn, P.A. and Fuerst, W. [1989]. *Introductory Digital Signal Processing,* John Wiley and Sons, Chichester, UK.

Martin, J.D. [1991]. *Signals and Processes,* Pitman Publishing, London.

Oppenheim, A.V. and Schafer, R.W. [1975 & 1988]. *Digital Signal Processing,* Prentice Hall, Englewood Cliffs, NJ.

Papamichalis, Panos (ed.) [1991]. *Digital Signal Processing Applications with the TMS320 Family, Volume 3,* Prentice-Hall, Englewood Cliffs, NJ.

Pratt, W.K. [1978]. *Digital Image Processing,* John Wiley and Sons, Chichester, UK.

Stafford, R.H. [1980]. *Digital Television, Bandwidth Reduction and Communication Aspects,* John Wiley and Sons, Toronto, Canada.

# 6

# Encoding of Waveforms – Increasing the Channel Bandwidth

Communications form a large part of modern life. Massive amounts of data are carried via telephone networks, TV channels and private and military data networks. The provision of data-carrying capacity is expensive and using it as efficiently as possible is paramount.

In this chapter we shall look at the different methods by which we encode speech and video signals so that we can transmit them over physical media. Unfortunately, in order to keep this chapter to a reasonable length we shall not be able to look at any forms of error management. Further reading of more in-depth textbooks, for example, Jayant and Noll [1984], will provide information on this topic.

This chapter is divided roughly into three parts. The first section covers the different forms of waveform encoding that can be used to reduce the bandwidth of either speech, image or data signals for transmission. The second section looks at vocoding schemes where the periodic nature of speech is used to design innovative encoding techniques. Finally, we shall look at encoding schemes that are specific to systems that combine image and speech signals, for example, videophones and other multimedia applications. Some of the coding methods mentioned are still under discussion by their respective standards committees at the time of writing this book.

## ANALOG WAVEFORM CODING

In our new digital world of DSPs we still come across some old-fashioned analog waveform encoding techniques. There are three types of analog pulse modulation techniques, as shown in Figure 6.1: amplitude, duration (width) and position. In pulse amplitude modulation (PAM) the amplitude of a train of pulses is modulated in accordance with the message signal. In pulse width modulation (PWM) the widths of the pulses are modulated. Finally, in pulse position modulation (PPM) the position relative to a mean is modulated. In general, the zero amplitude message signal is represented by a nonzero value of pulse to ensure there are no missing pulses. This maintains a constant pulse rate allowing simpler demodulator design.

We came across one application of analog pulse modulation in Chapter 3. PWM and PDM systems are found in bit stream DACs. Pulse density modulation (PDM) is a combination of PWM and PPM. Bit stream DACs use these techniques because it is possible to recover the message signal by using a simple switched capacitor circuit.

## DIGITAL WAVEFORM CODING – PULSE CODED MODULATION (PCM)

In digital waveform coding we modulate a train of pulses in accordance with a digital incoming message signal. The most common form of digital waveform coding is known as Pulse Coded Modulation (PCM), where the output from the coder is a serial train of equally sized pulses chosen to represent the message signal as efficiently as possible.

PCM is actually a general description for the serial data stream produced by digitizing an analog signal (Figure 6.2). There are many variations of the basic PCM scheme which are enhanced for use in specific applications. For example, we came across $\mu$-law and A-law PCM in Chapter 3. These are generally used for telephone systems.

Once we have used PCM to encode our signal we are still left with a baseband signal with significant dc and low-frequency components. This means it is unsuitable for transmission over any appreciable distance. Therefore, a PCM output is usually modulated onto a higher-frequency carrier wave using some form of digital continuous wave modulation, for example, amplitude shift keying (ASK), phase

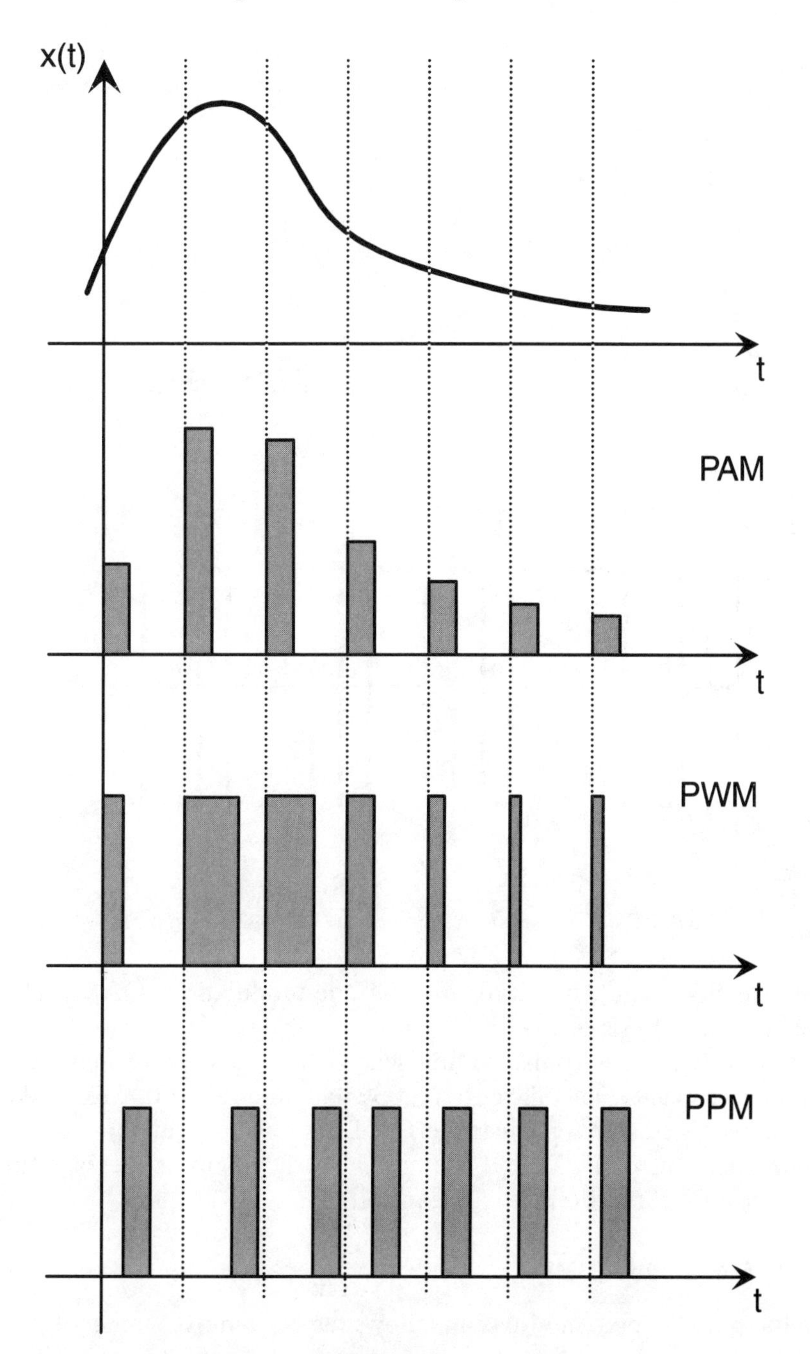


**FIGURE 6.1.** Analog pulse modulation in amplitude (PAM), width (PWM) and position (PPM)

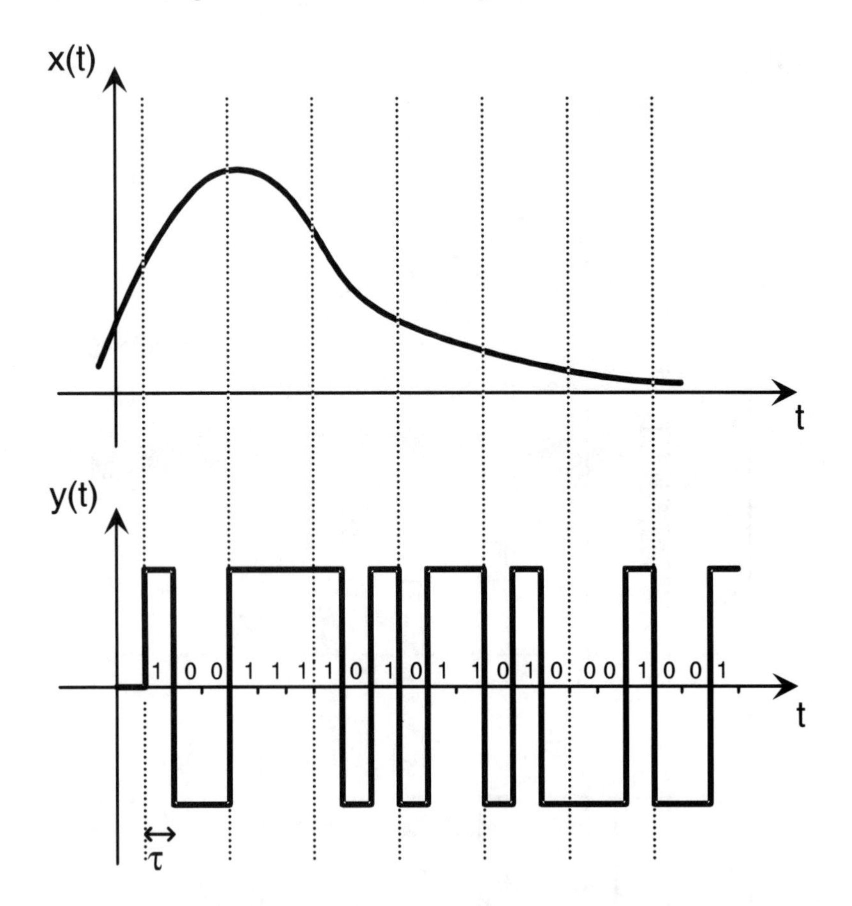


**FIGURE 6.2.** Pulse coded modulation (PCM), 3-bit example

shift keying (PSK), quadrature amplitude modulation (QAM), etc. (see Carlson [1981], Clark [1983]).

Generally, in any pulse coding scheme we must try to make the channel capacity as efficient as possible and also try to make the transmitted signal robust to error. Let's now look at some of the different forms of PCM and examine how they help us increase the amount of information we can transmit.

### Delta Modulation (DM)

Delta pulse coded modulation allows us to remove some of the redundant bits in a PCM data stream and is by nature less sensitive to channel errors. We briefly looked at delta modulation in Chapter 3 with reference to sigma delta ADCs. This is an extension of PCM that

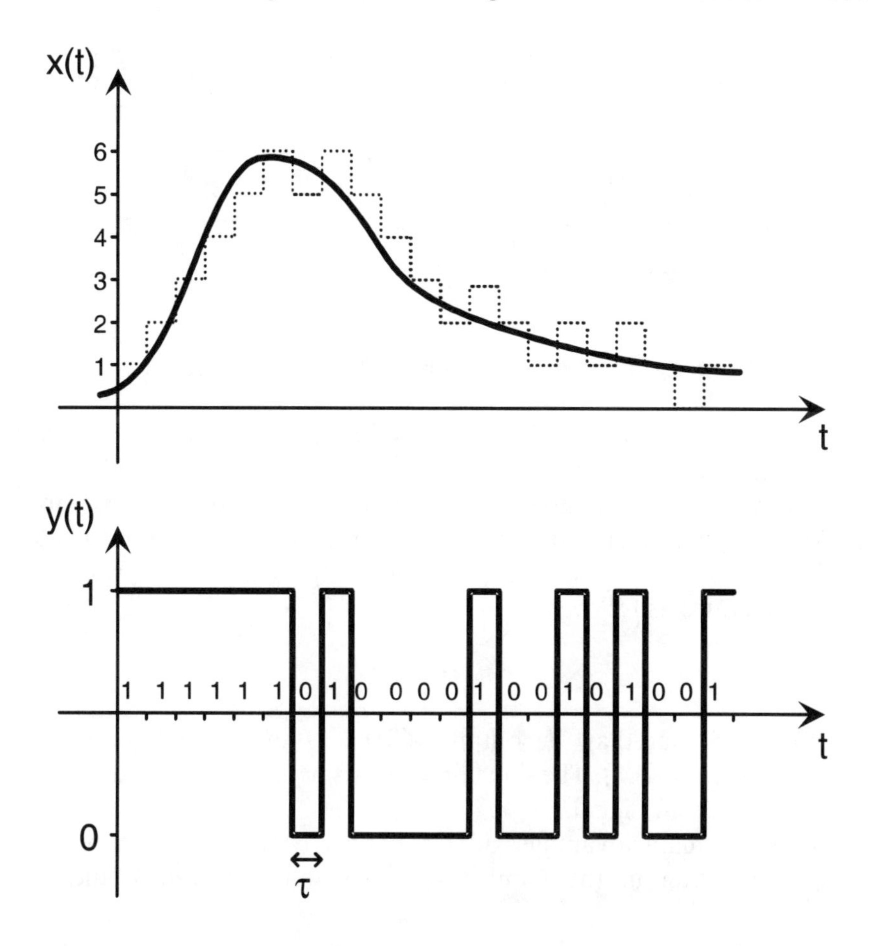


**FIGURE 6.3.** Delta modulation (DM)

quantizes the analog signal using only one bit. The implementation of delta modulation is simple, as we can see in Figure 6.3.

If the input signal is larger than the accumulated digital value at a sampling point then the digital value is incremented by one. Equally, if the input is less than the accumulated digital value at a sampling point, the digital value is reduced by one. The output word for each sampling period is just a one or a zero depending on whether we increased or decreased our running total. This implies that to transmit the information on the same signal we can use $\frac{1}{n}$ times the frequency for DM that we require for PCM (where $n$ is the number of PCM bits). The block diagram of a delta modulator and demodulator is shown in Figure 6.4.

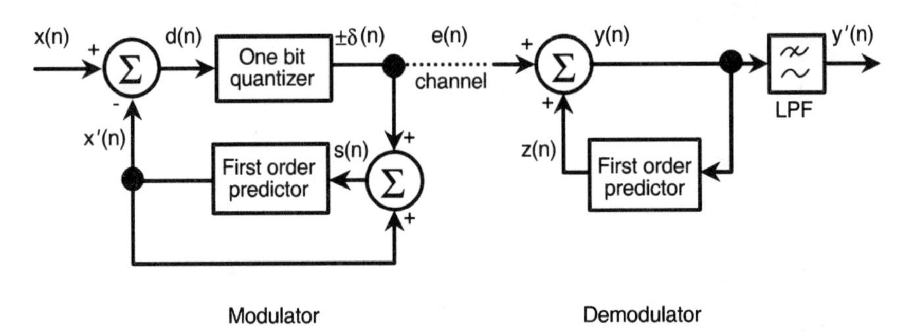


FIGURE 6.4. Block diagram of delta modulator and demodulator

Before being input to the delta modulator, the original analog waveform is passed through a sample and hold circuit to produce a discrete waveform, as shown in Figure 6.5. At $t = 0$, assuming no previous inputs, $x(n)$ is quantized to 1 or 0. The feedback element produces a prediction of the expected next input value. If we assume that it operates as follows:

$$x(n) = s(n)$$

we can work out that the output will be as shown in Figure 6.6 (assuming $\delta(n) = \pm 1$). The same predictor is used in the demodulator to reconstruct the signal.

One of the main advantages of DM that can be seen from Figure 6.6 is the reduction in the number of transitions between 1 and 0

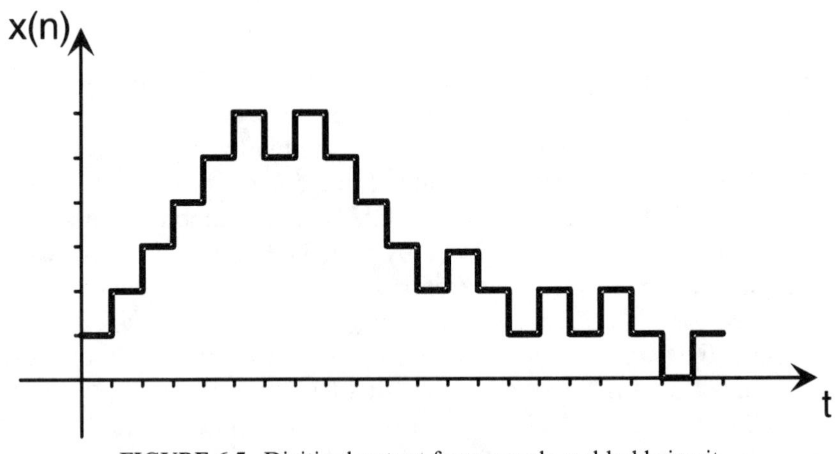


FIGURE 6.5. Digitized output from sample and hold circuit

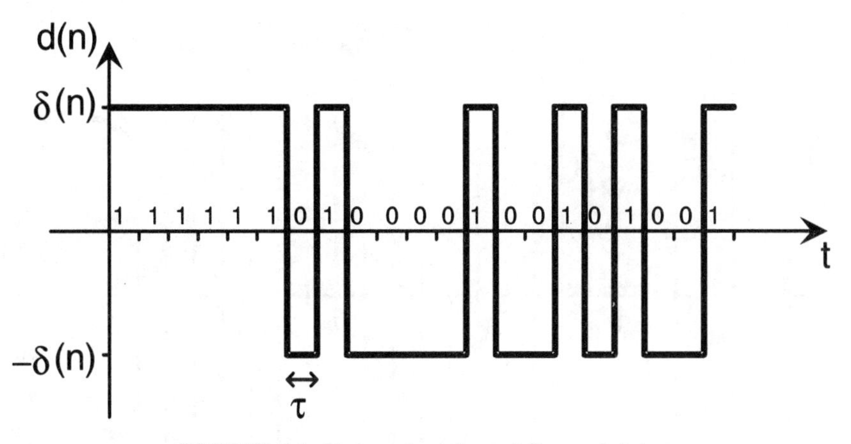


**FIGURE 6.6.** Data output from delta modulator

compared with the PCM example of Figure 6.2. Our PCM case had the same bit rate ($\tau$) but we need three bits to represent each sample. This means that the effective sample rate is only one-third of the bit rate. In the case of DM we have one bit per sample and so for the same output bit rate we can have three times the number of samples and also less transitions.

Fixed step-size DM is often referred to as linear DM (LDM). There are two major causes of error in LDM – slope overload and granularity. If the step size, $\delta(n)$, is too small we get slope overload, as shown in Figure 6.7. If $\delta(n)$ is too large we suffer from granularity, as shown in Figure 6.8. It is usually necessary to sample the signal at a

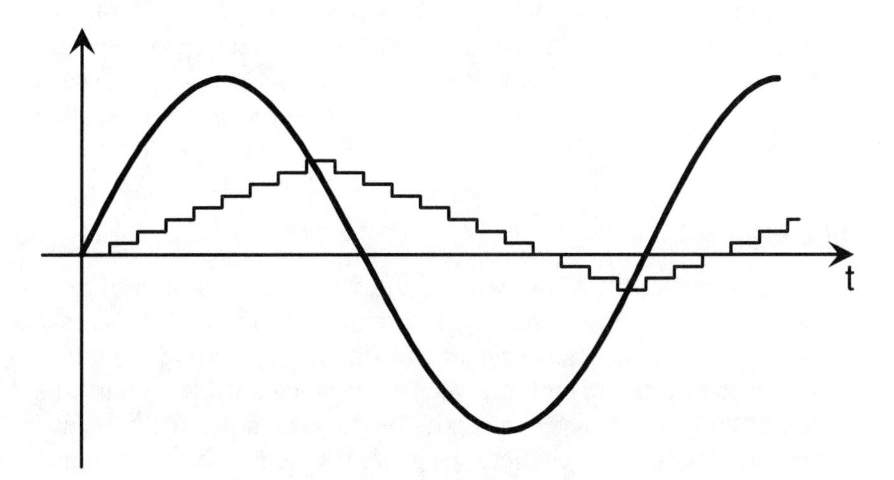


**FIGURE 6.7.** Delta modulation with slope overload due to too small a step size

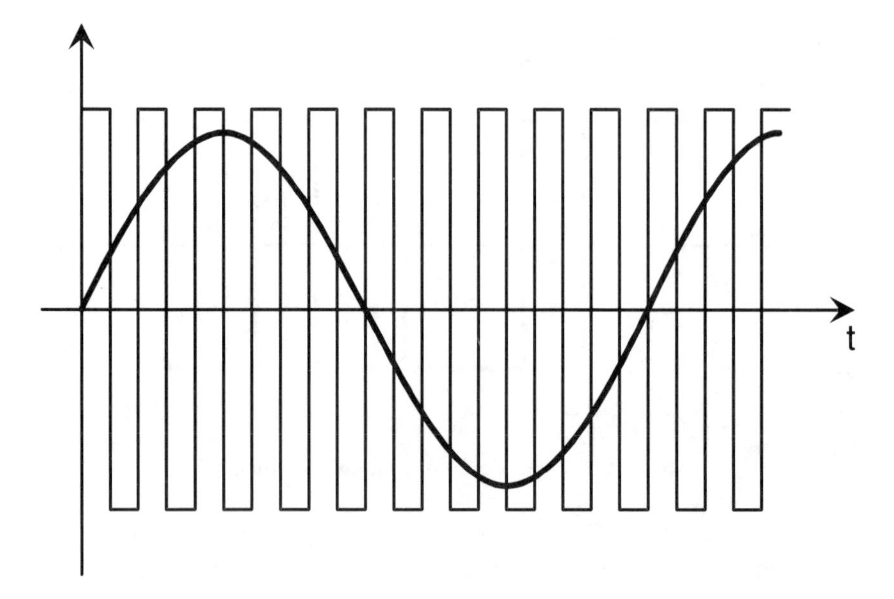


**FIGURE 6.8.** Delta modulation with granularity due to too great a step size

faster rate for DM than for PCM to avoid these two problems and much of the perceived advantage is lost.

Consequently, a great deal of care is necessary when choosing a value for $\delta$. If we know the characteristics of the incoming waveform this will not be a problem, but if the waveform is less predictable we have to consider another approach. An example of a waveform where linear DM is often inappropriate is human speech. There will be some very large variations in amplitude between silent sounds ("ph" or "sh") and loud sounds ("t" or "a"). This may give us a problem with particular words and certainly across the full range of speech. In such cases adaptive predictors are used, which alter the step size depending on a number of previous samples. We shall look at these shortly.

### Differential PCM (DPCM)

Differential pulse code modulation (DPCM) is an extension of delta modulation. It again utilizes the redundancy in analog signals, especially speech and image signals. In differential coding the difference between a discrete input signal and a predicted value is quantized to one of "p" values. The complexity of the system is directly related to the complexity of the predictor. A general system is shown in Figure 6.9.

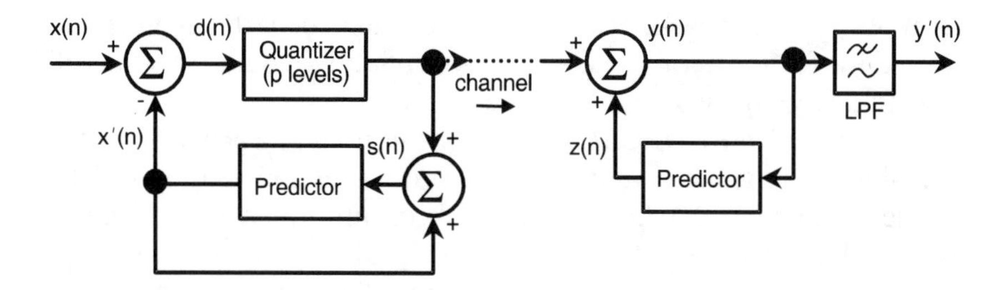


**FIGURE 6.9.** Differential PCM modulator and demodulator

The DPCM predictor can generally be expressed by the following equation:

$$x'(n) = \sum_{R=1}^{N} h_R s(n - R)$$

where $h_R$ is the set of predictor coefficients. Note that if $h_1 = 1$ and all other coefficients are zero, we have our delta modulator.

It is also interesting to note that the general predictor equation is actually the same as the equations we used in Chapter 4 for FIR filters. This is why in many textbooks you will see the DPCM system drawn with a digital filter in the feedback loop rather than a predictor. In addition, since the predictor is simply a filter we know that a DSP will provide an ideal solution.

The advantage of using DPCM over PCM is the gain in signal-to-noise ratio (SNR), i.e., the improved quality of the output waveform. As DPCM is only quantizing the difference between signals and not the absolute value, it will incur far less quantization error despite only using the same number of bits as a PCM system. DPCM is therefore much more accurate for the same system resources.

The relative merits of different order predictors are discussed in depth in Jayant and Noll [1984]. In general, we can say that the gain in SNR saturates at approximately $N = 2$ for DPCM speech and at around $N = 3$ for intraframe image processing (Habibi [1971]).

## Adaptive DPCM (ADPCM)

The quality of the coding algorithm depends on our knowledge of the signal statistics. If we have a clearly defined input signal we can design time invariant (i.e., fixed) predictors. In most cases, although the long-term statistics are well understood, the signal departs significantly from these for shorter periods of time. An adaptive coding scheme will give an advantage where this is true.

The term *adaptive DPCM*, or ADPCM, is used as a general title for two different schemes – adaption of the quantizer and adaption of the predictor. Figure 6.10 shows an example of uniform step-size quantization adaption depending on the input signal amplitude range. Step-size estimation can be performed in one of two ways, forward or backward estimation (DPCM-AQF and DPCM-AQB, respectively). Block diagrams for each are shown in Figure 6.11.

In forward estimation (AQF), the input samples are buffered and used to estimate the level of the input before coding, thereby reducing the effects of quantization noise. AQF suffers from the need to transmit both the coded signal and the level information. For this reason AQB is more common. Another drawback that renders AQF unusable in certain applications is the delay in signal transmission due to coding. For example, if we consider speech to have a bandwidth of 4kHz and we sample at the Nyquist rate (8k samples/sec), an AQF algorithm using 256 values for level estimation would cause a delay of 16ms before any output was produced.

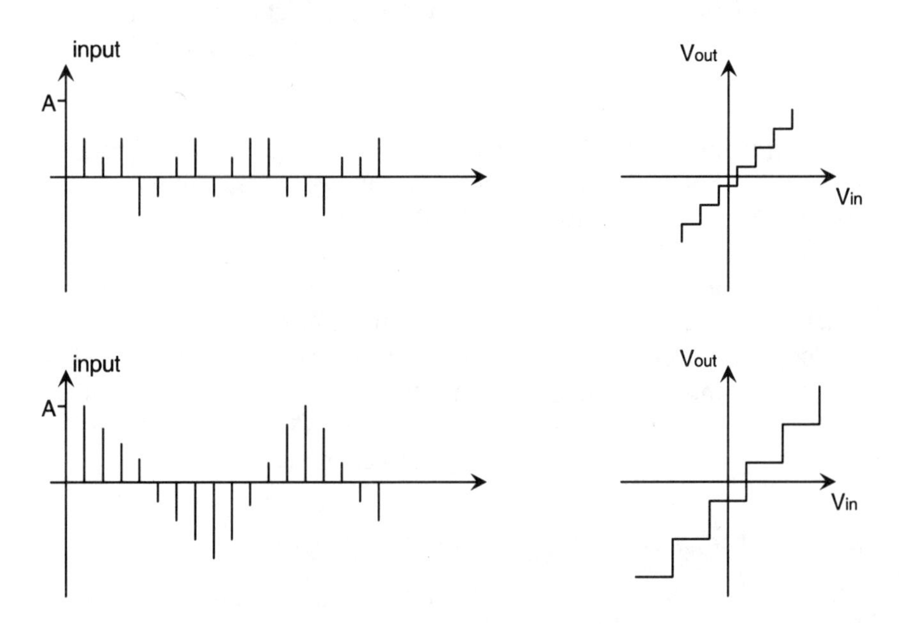


**FIGURE 6.10.** Example of adaptive quantization step size – small step size for low amplitude signal, large step size for high-amplitude signal

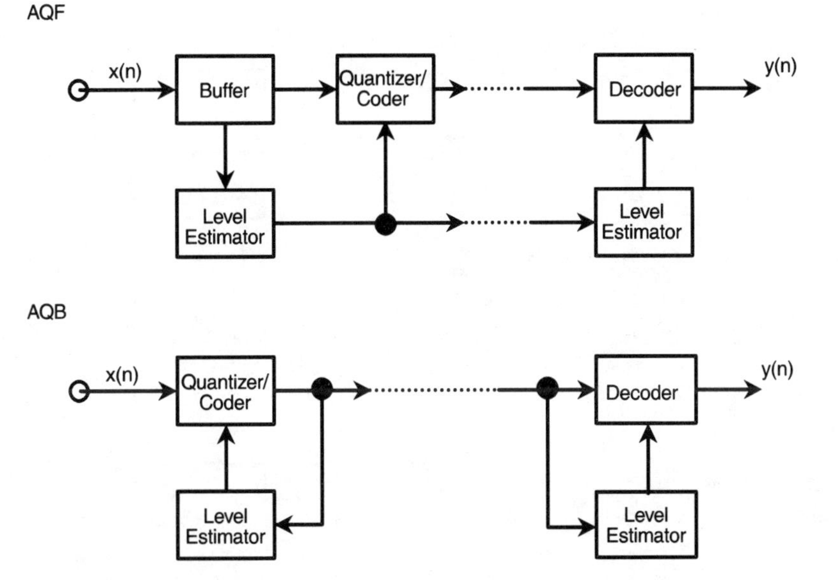


**FIGURE 6.11.** DPCM-AQF and DPCM-AQB modulation and demodulation – note buffer on input of AQF modulator

Adaptive quantizers offer an improvement in SNR of between 3dB and 7dB over nonadaptive schemes, even with 3-bit speech-coding algorithms. The quality of the adaption again depends on the quality of the estimator, which could track the mean of the incoming signal or possibly its variance. Again, we are faced with a trade-off of complexity against system realization. The performance of the DSP chosen directly affects the sophistication of the estimation scheme used. DPCM-AQF and AQB schemes are suitable for applications with bit rates of around 32kbit/s.

The most common approach to speech-coding systems is to combine adaptive quantizers and adaptive predictors. With adaptive prediction systems we can increase the gain in SNR significantly with $R = 10$, compared to $R = 2$ for nonadaptive. It is this form of adaptive DPCM system that we shall from now on refer to by the acronym ADPCM.

In ADPCM we have the choice of making the adaption process either feedforward or feedback. Since we are adapting both quantizer and predictor there are four possible combinations. The CCITT standard G.721 employs feedback adaption of the predictor and feedback adaption of the quantizer (Figure 6.12).

The input to the encoder is the standard CCITT 64kbit/s PCM

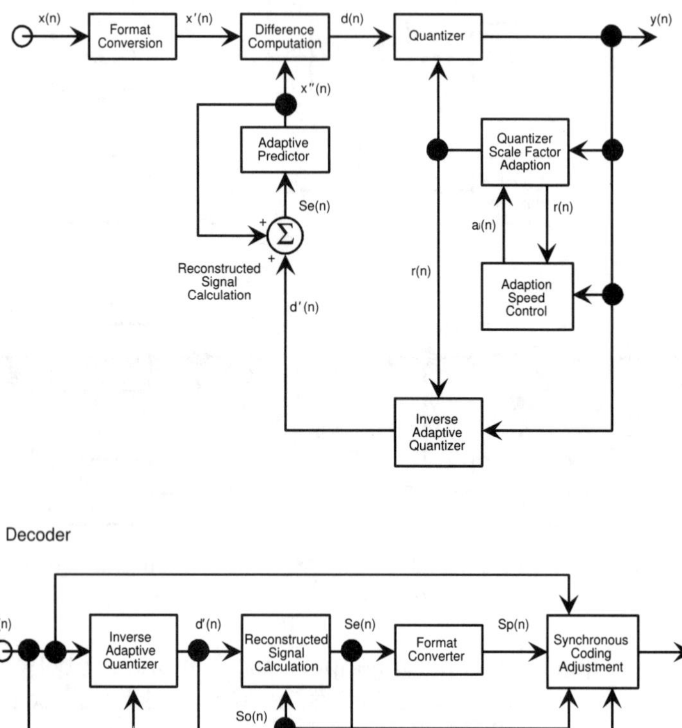


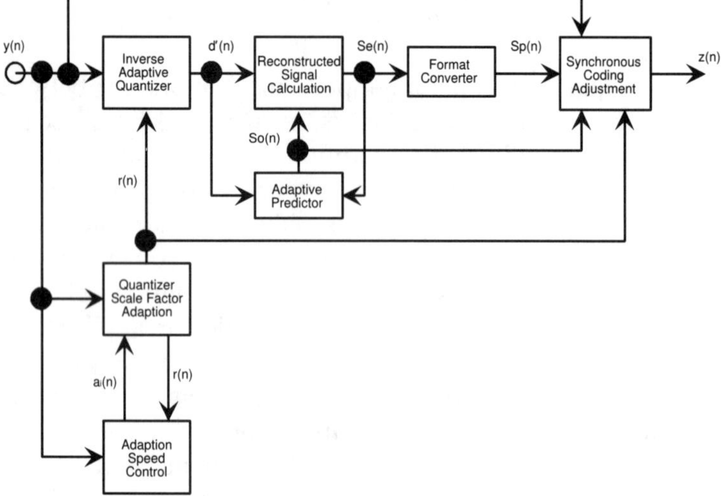


**FIGURE 6.12.** CCITT G.721 ADPCM modulator (encoder) and demodulator (decoder)

coded speech. The CCITT standard uses a companding algorithm, either A-law or $\mu$-law depending on whether it is for Europe or the United States, respectively. The encoder transforms the PCM digital data stream into ADPCM at 32kbit/s and hence is usually referred to as a transcoder. The format conversion box in the encoder reconstructs a linear PCM data stream from the incoming companded version. Conversely, in the decoder, this box performs companding on the reassembled signal.

A 4-bit adaptive quantizer is used on the difference signal. Note that as we have feedback adaption of the predictor, the encoder also contains an inverse adaptive quantizer to reconstruct the differential signal values. The predictor then uses previous estimated input values, $s_e(n)$, to calculate the next value which is fed back and subtracted from the input to form the differential input signal, $d(n)$.

One of the advantages of using feedback predictors is that the same predictor block is used in both the encoder and decoder sections, simplifying the design. The same is true for the adaptive quantizer. Note that there is an extra block included, referred to as adaption speed control. This allows extra control of the quantizer with fast adaption for large amplitude variations (e.g., speech) and slower adaption for signals that vary more slowly (e.g., data).

This appears to be a very complicated function. So, in order to put it into some perspective with respect to DSPs, let's look at its implementation. The predictor algorithm consists of a sixth-order FIR filter and a second-order IIR filter and all eight coefficients are adapted using a gradient equation on the differential input signal. A half-duplex G.721 system is available as a mask-programmed first-generation TMS320 called the TMS320SA32. A full description of the circuit can be found in Charbonnier et al. [1985], but we shall briefly describe the circuit here.

A block diagram of the system is shown in Figure 6.13. Data is read every 125$\mu$s from the codec interface and the corresponding ADPCM

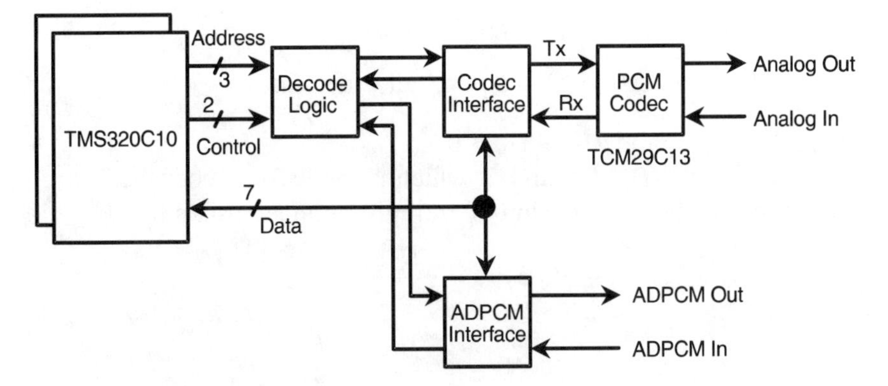


**FIGURE 6.13.** G.721 implementation using TI TMS320C10 DSP

is available after 53$\mu$s. The DSP has a 160ns cycle time. Therefore, we can see that it actually takes approximately 330 machine cycles to complete the mathematics. Less than 1.5k words of program space are required so that all the code can be mask programmed onto the DSP, minimizing the number of integrated circuits.

The masked TMS320SA32 transcoder offers an extremely cost-efficient solution for applications such as voice mail, which only require half-duplex operation. Obviously, for full-duplex operation we could use two masked devices. Alternatively, we can implement a full-duplex CCITT ADPCM system on a single second-generation TMS320C25.

### Adaptive Delta Modulation (ADM)

We shall not spend any time on this technique since we have covered the basics in the section on ADPCM. ADM generally uses feedback quantizer adaption to avoid having to transmit the extra level information. It has the advantage that it is simpler to implement than ADPCM.

### Continuously Variable Slope Delta Modulation (CVSD)

Although ADM is simple to implement it suffers significant degradation in speech quality if there are errors in transmission. These errors can propagate through the speech for a considerable amount of time. To recover from these errors it is necessary to introduce some "leakage" into the predictor and quantizer.

In CVSD the step-size adaption depends on the two previous values of the encoder's output signal, $y(n)$. The step size $\Delta(n)$ is given by:

$$\Delta(n) = \beta\Delta(n-1) + D_2 \quad \text{if} \quad y(n) = y(n-1) + y(n-2)$$

$$\text{or} \quad \Delta(n) = \beta\Delta(n-1) + D_1 \quad \text{in all other cases}$$

where $0 < \beta < 1$ and $D_2 \gg D_1 > 0$.

The values for $D_1$, $D_2$, and $\beta$, which is the leakage coefficient, are related to the required maximum and minimum step sizes by:

$$\Delta_{max} = \frac{D_2}{1-\beta}$$

$$\Delta_{min} = \frac{D_1}{1-\beta}$$

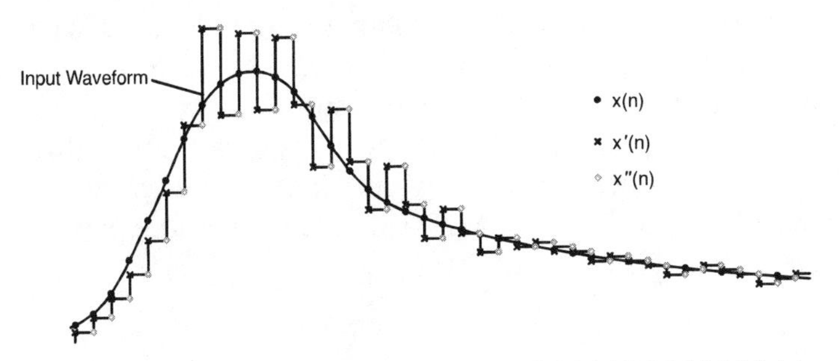


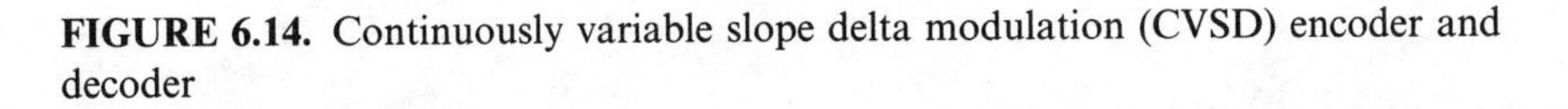


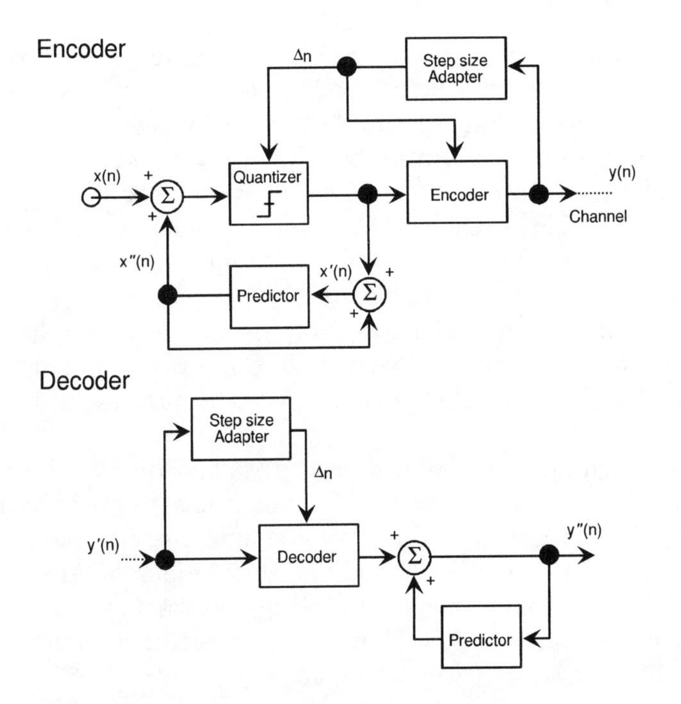


**FIGURE 6.14.** Continuously variable slope delta modulation (CVSD) encoder and decoder

The values for these three coefficients are chosen carefully to ensure that the CVSD coder can follow the incoming message signal and are generally derived from long-term signal statistics. The first set of equations shows that the step size increases when we have continuous runs of either ones or zeroes. Looking at Figure 6.14 we can see that this is exactly what is required when we are suffering from slope overload. In all other cases the step size decreases. If we choose $\beta$ to be almost 1, the rate of change of $(\Delta n)$ will be slow. Conversely, if we choose $\beta$ to be close to 0, adaption will be fast. In general $\beta$ is chosen first. Then $D_1$ and $D_2$ are calculated using the $\Delta_{max}$ and $\Delta_{min}$ equations.

Although CVSD produces poorer-quality speech than ADM, APC or ADPCM, it is much less sensitive to transmission errors. CVSD was quite widely used until the ADPCM schemes became popular and were then standardized by the CCITT.

### Adaptive Predictive Coding (APC)

If we concentrate for now on speech coding, the next class of algorithm is adaptive predictive coding (APC). APC is slightly different from the coding schemes we have looked at so far. It considers the speech waveform to be repetitive with a period significantly greater than the average frequency content. The signal is then split into high- and low-frequency parts and two prediction algorithms are used. The high frequency is estimated using a spectral predictor and the low frequency by a pitch predictor.

The spectral predictor usually is of fourth order, deriving its coefficients from the input signal in the same way as we saw for ADPCM. The input to the pitch predictor is the output of the spectral predictor.

APC has been used for coding speech at 9.6 and 16kbit/s. Its main drawback is quality. This is why CELP or multipulse implementations are preferred. They give better quality for comparable complexity (see later). Nevertheless, there are many implementations of APC, as it works well in noisy environments. Further analysis and references can be found in Papamichalis [1987] and an example of the implementation of an APC circuit can be found in Lin (ed.) [1987].

### Subband Coding (SBC)

Subband coding is a general term that can use any of the previous PCM-based techniques. Basically, the speech waveform is first split into subbands using band-pass filters and then each subband is encoded using either ADM, ADPCM, APC or any other technique (see Figure 6.15).

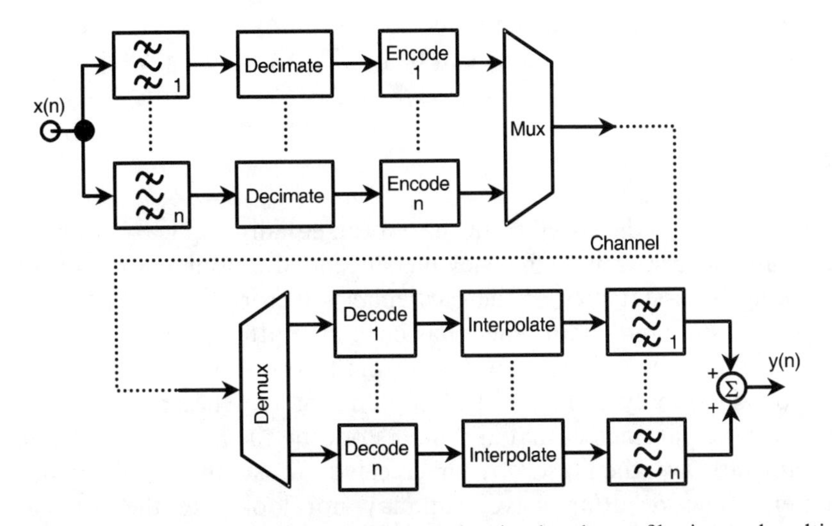


FIGURE 6.15. Subband coder and decoder showing band-pass filtering and multiplexing

In splitting the speech into subbands, any quantization noise is kept within that band and cannot interfere with any other subband. Also, the available bits can be allocated between bands according to perceptual criteria, enhancing the speech quality as perceived by the listeners, though not necessarily improving the SNR.

The encoder section includes decimation circuits that allow the control of the number of bits per subband. In SBC, decimation simply refers to throwing away unwanted samples. For example, assuming we have a fixed number of bits to represent the input signal, and there are two equally important subbands, half the bits will be assigned to each. When we then sample both subbands at the original sampling frequency, we obtain twice the permitted number of bits and must therefore discard every other sample in each subband. After encoding, the two channels are simply multiplexed for transmission.

At the decoder, interpolation is used to re-create the missing samples before the subbands are filtered and summed to produce the reconstructed signal.

At bit rates of between 9.6kbit/s and 32kbit/s, subband coding with APC compares favorably with ADPCM and ADM. However, the complexity of the system may be higher depending on the number of subbands. In addition, the design of the band-pass filters must be carefully controlled to avoid interference. Special filters known as quadrature mirror filters (QMF) are generally used (Esteban and

Galand [1977]). As with the other coders, an example of the implementation of subband coding can be found in Lin (ed.)[1987].

## VOCODERS

All the methods discussed so far have been equally applicable to both data and speech. Their efficiency has depended on our knowledge of the long-term statistics of the particular waveform, but apart from APC and subband coding there has been little attempt to understand the nature of the waveform we are trying to transmit.

Vocoders analyze the spectral contents of the speech to try and identify the parameters that are understood by the human ear. These parameters are then used at the receiver to synthesize the voice pattern. The resulting waveform may not look like the original speech signal, but the differences are not discernible by our ears. This is very different from the previous waveform coders where we tried to ensure the signal was accurately reproduced at the receiver output.

Vocoders have become popular because they ensure reasonable representation of the speech waveform at low bit rates – from 2.4kbit/s to 9.6kbit/s. So, where we have limited bandwidth, or are limited in how much information can be stored at any one time, vocoding techniques ensure the best use of the available resources.

Parametric vocoders split the speech signal into a model of the

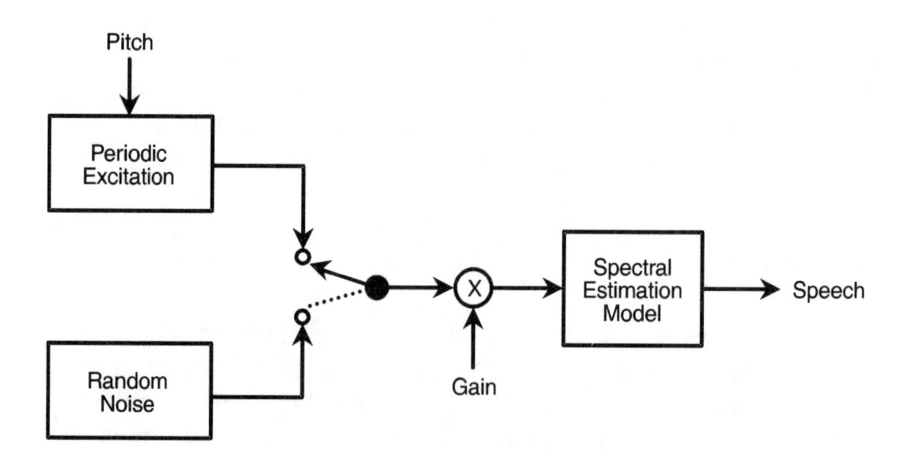


**FIGURE 6.16.** The basic blocks of a generic vocoder

spectral envelope and an excitation signal (Figure 6.16). These are then coded separately as they are assumed to be independent. It is simplest to think of this in relation to the human speech production system – the excitation signals are produced by the vibration of the vocal cords over a wide range of relatively high frequencies, referred to as the pitch of the speech. The spectral envelope model is the shape of the mouth, which forms the sound of the words we speak. There is also a random noise generator included to represent the production of "unvoiced" sounds, for example "s" or "f".

There are many different types of vocoders. We shall attempt to outline the main features of the most popular types in this book along with the differences between them. References to further reading can be found at the end of the chapter. Before we start the discussion on vocoders, this is an appropriate time to look at windowing since all vocoders work on only a short section of speech.

## Windowing

Although speech is a continuously varying and almost random waveform, vocoders exploit the fact that for short periods of time the signal appears to conform to a pattern. It is this pattern that they attempt to extract in order to produce a digital representation of the speech. So the first stage in any vocoder is the windowing of the region of interest of the input signal (Figure 6.17).

Let's first look at the effects of using a rectangular window, as

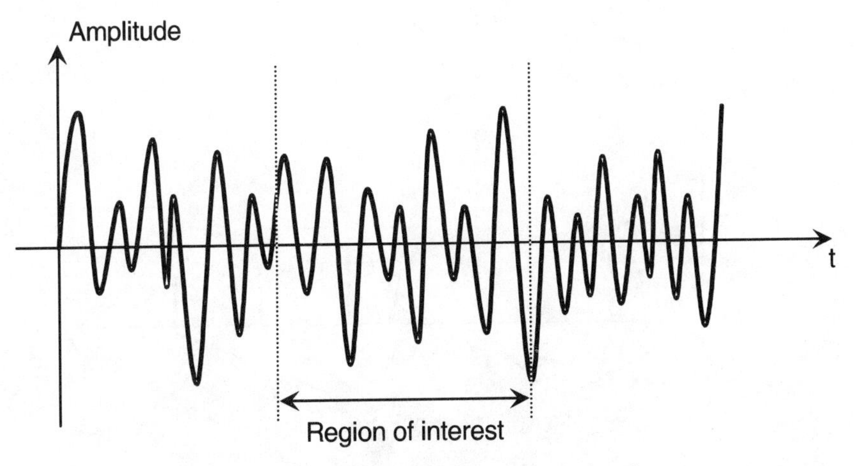


**FIGURE 6.17.** Windowing a speech signal.

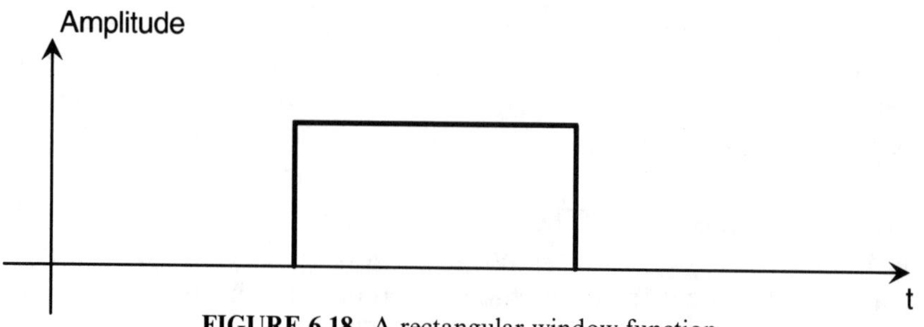


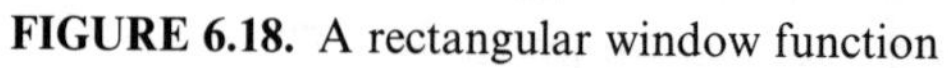
**FIGURE 6.18.** A rectangular window function

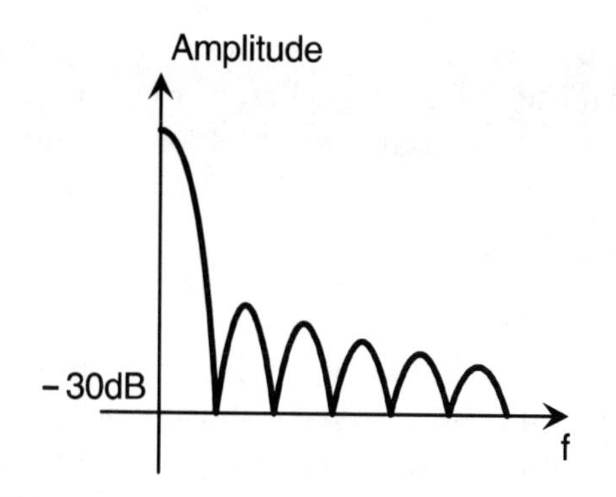


**FIGURE 6.19.** Frequency spectrum of a rectangular window

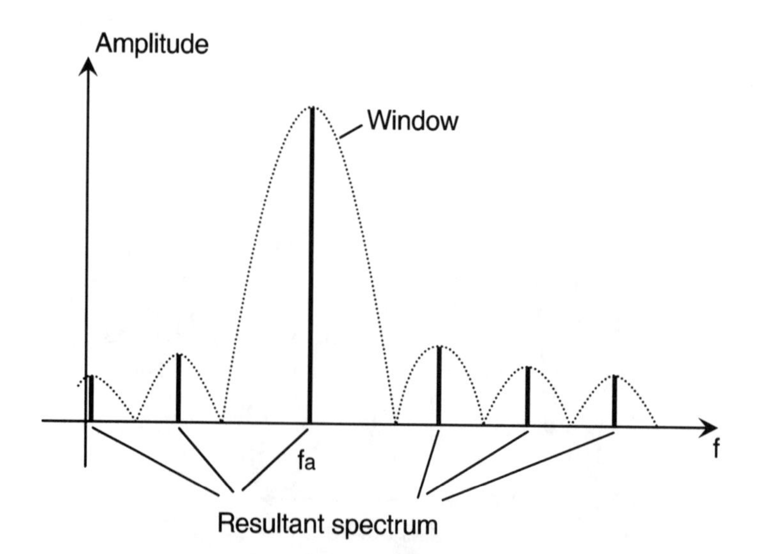


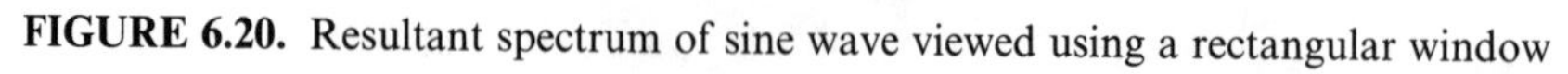
**FIGURE 6.20.** Resultant spectrum of sine wave viewed using a rectangular window

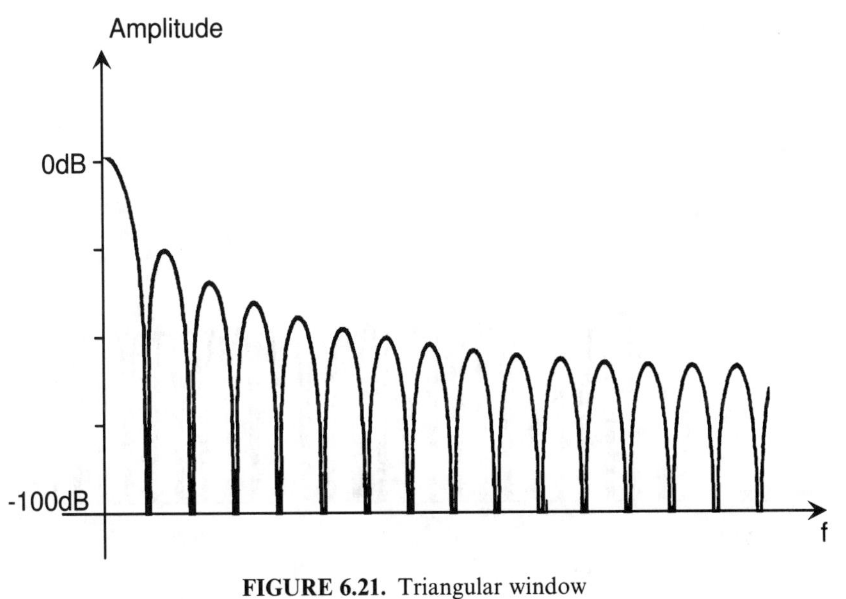


**FIGURE 6.21.** Triangular window

shown in Figure 6.18. The resulting spectrum is a combination of the spectrum of the window and the speech signal. The window's spectrum is shown in Figure 6.19 and is generally known as the sinc function ($\{\sin x\}/x$). For simplicity, consider the input signal to have

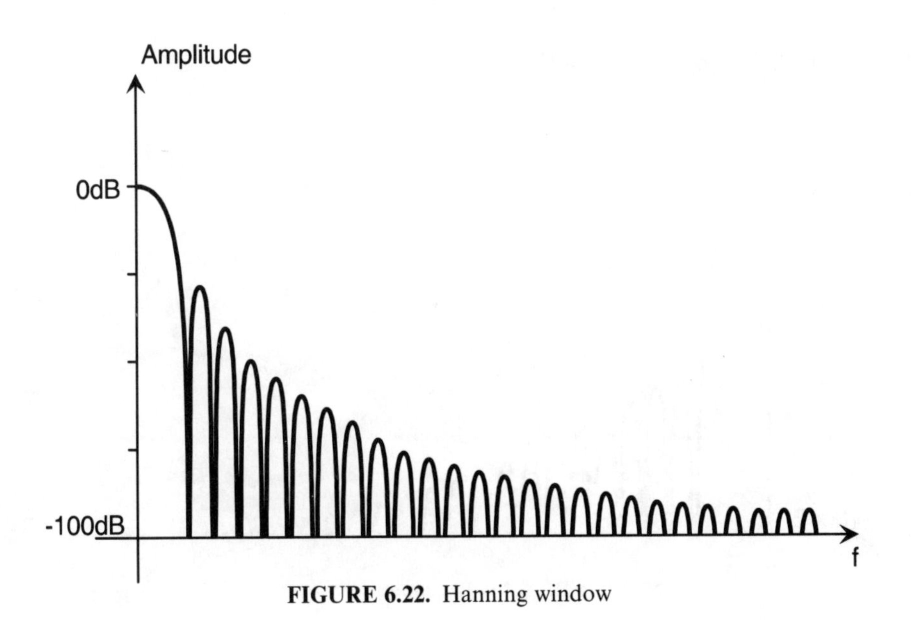


**FIGURE 6.22.** Hanning window

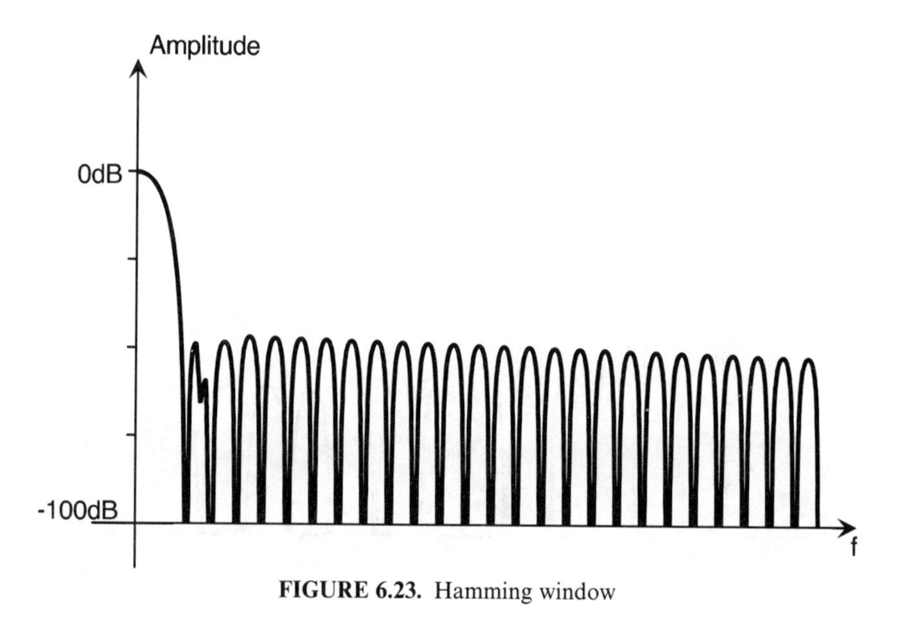


**FIGURE 6.23.** Hamming window

a single-frequency at $f_a$ (a sine wave). The resultant spectrum using this rectangular window will look like Figure 6.20. What should have been a single frequency component has now been repeated many times. If we demodulated this spectrum the resulting waveform would obviously not be a pure sine wave.

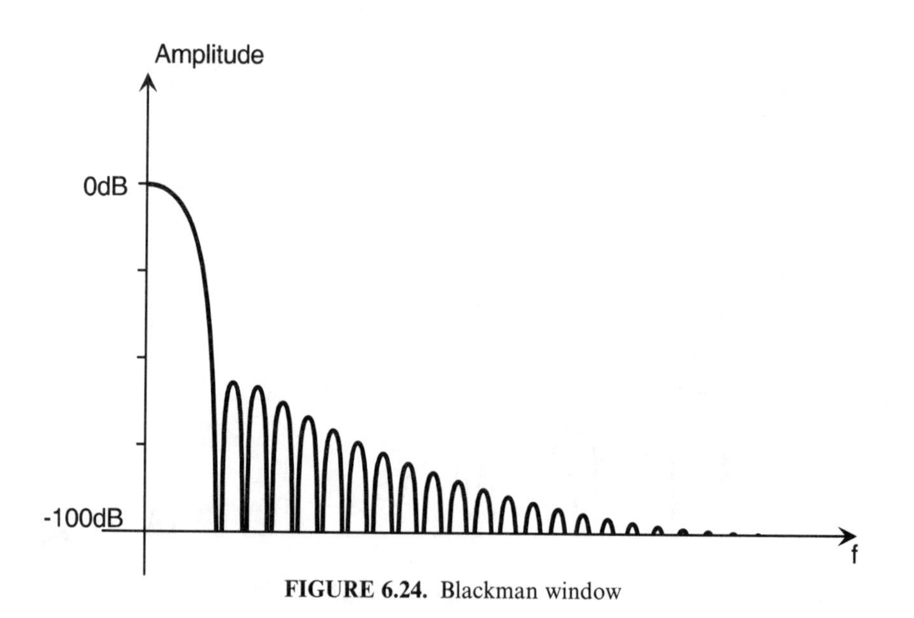


**FIGURE 6.24.** Blackman window

From this simple example we can see that windowing a continuous time signal affects the spectrum. In fact, a rectangular waveform is the worst possible example and is rarely used in practice. If we stick to the example of a single frequency we can see that an ideal window would have a rectangular function in the frequency domain, which would require a sinc function in the time domain. This is quite a difficult function to achieve and a compromise is invariably used.

Figures 6.21 to 6.24 show the most popular types of window that offer a trade-off between the width of the main lobe and the roll-off of the sidebands. Windowing is used in almost all digital signal processing since it is only possible to consider a portion of the incoming signal at one time. In digital filtering, the windowing function is built into the calculation of the filter coefficients, as explained by Bateman & Yates [1988]. More detailed information on windowing can also be found in Martin [1991].

## Channel Vocoder

The oldest and most widely studied vocoder was conceived by H. Dudley in the 1930s and demonstrated at the New York Fair in 1939. The channel vocoder exploits the fact that the ear is relatively insensitive to the phase of the speech signal over short periods of 10–40ms, so only the magnitude of the speech needs to be transmitted.

The analyzer uses band-pass filters to split the speech into subbands (Figure 6.25). The bandwidths of the filters are designed to increase

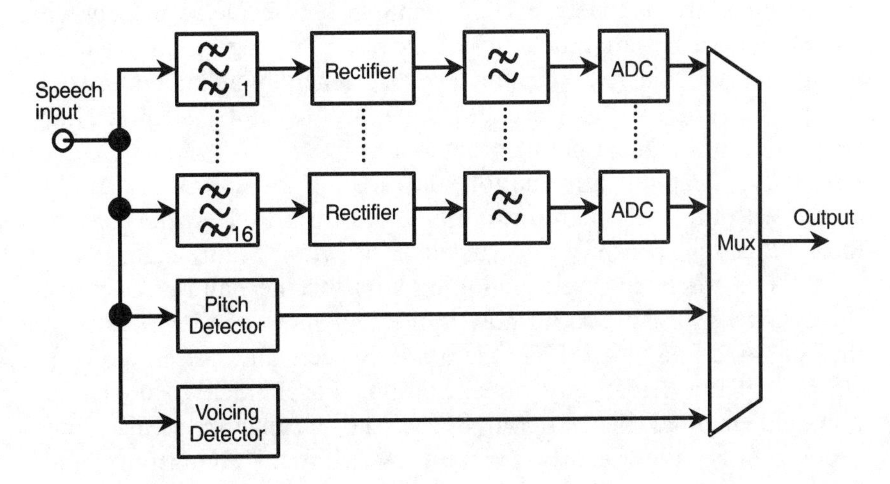


**FIGURE 6.25.** Channel vocoder – analysis

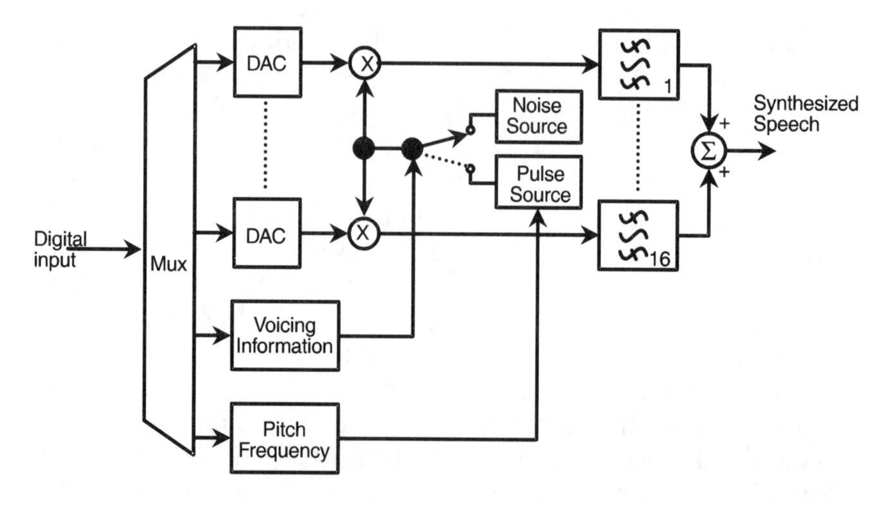


**FIGURE 6.26.** Channel vocoder – synthesis

proportionally with frequency to match the response of the human ear. Each band is rectified and low-pass filtered in order to determine its envelope, and then this is digitized and encoded in some way. The sampling is typically performed every 20ms.

The speech signal must also be analyzed for the pitch of the signal during the window. The pitch frequency can tell us whether the speech is voiced or unvoiced. This is the crucial part of a vocoder and numerous pitch-detection algorithms have been devised over the years. Voiced segments of speech with clear periodicity and unvoiced segments that are not periodic are relatively easy to identify. The segments in between these two extremes are much more difficult and no algorithm has been found that is considered exceptional by all listeners. We shall not discuss any pitch estimators in this book. Papamichalis [1987] offers a clear description of the most popular methods.

Finally, the pitch information and voicing indicators are multiplexed with the subband information and transmitted. At the receiver the synthesized speech is constructed as shown in Figure 6.26.

The first vocoders used analog circuits for the band-pass filters. With today's DSPs, we can now implement most of the vocoder in the digital domain. In 1966 the Joint Speech and Research Unit (JSRU) of the British government designed a channel vocoder that was implemented by Marconi Space and Defence Systems. This vocoder was subsequently digitized by Marconi (Kingsbury and Amos [1980]).

## Homomorphic Vocoder

The homomorphic vocoder is based on the assumption that the short-term speech signal consists of a slowly varying vocal tract spectrum and a higher-frequency excitation spectrum. The complete signal consists of the modulation of the excitation spectrum by the vocal tract spectrum. If we consider the log of the frequency spectrum, only the addition of the two components is required, rather than the multiplication (Figure 6.27).

If we assume the spectrum shown in Figure 6.27 is finite and pretend that the two components are functions of time, we could derive a spectrum for the waveform, as shown in Figure 6.28. This is

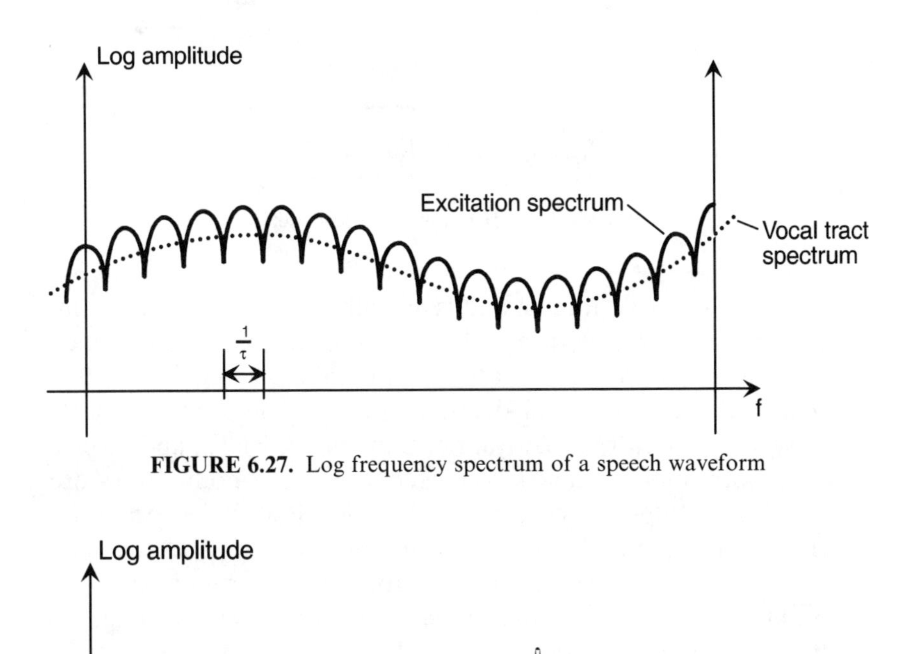


**FIGURE 6.27.** Log frequency spectrum of a speech waveform

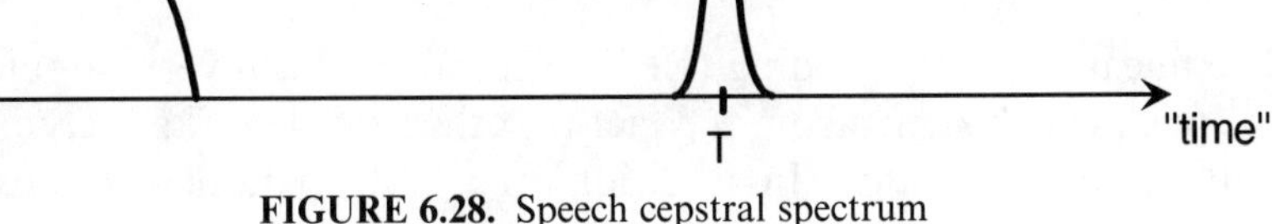


**FIGURE 6.28.** Speech cepstral spectrum

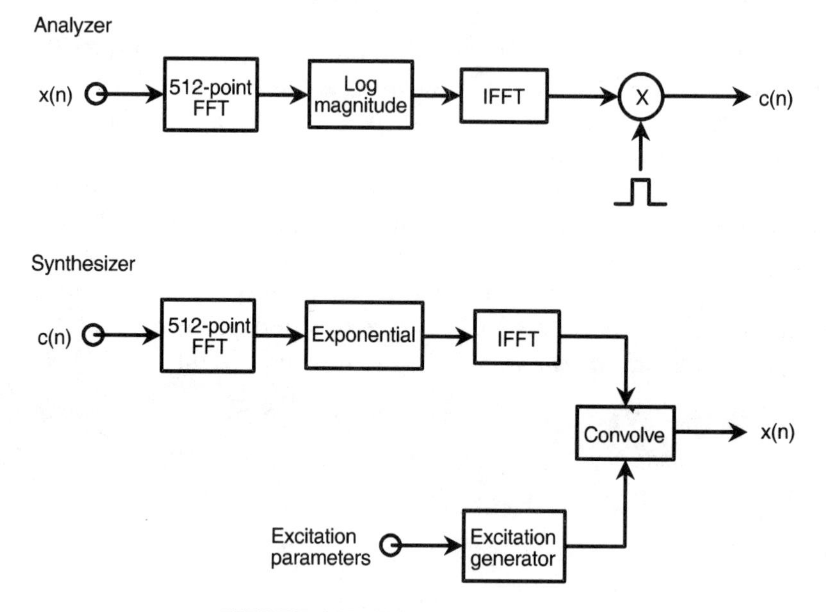


**FIGURE 6.29.** Homomorphic vocoder

known as the cepstral domain and is actually a kind of time domain. We can see that in the cepstral domain with a sensible cut-off "time" we can easily separate the excitation and vocal tract signals.

Figure 6.29 outlines a block diagram of the homomorphic vocoder. We can see that it utilizes fast Fourier transforms (FFTs) and inverse FFTs in both analyzer and synthesizer, where 512-point FFTs are most common. After the cepstrum has been windowed, the vocal tract information is separated from the excitation spectrum and encoded. In general, the excitation information from the cepstrum is not used but is obtained separately using a pitch estimator. After encoding, the output from the pitch estimator is multiplexed with the vocal tract information for transmission. At the receiver, or synthesizer, the reverse process is implemented, and the final synthesized speech is obtained by convolving the two sets of data.

### Linear Predictive Coding (LPC)

The linear predictive coding (LPC) method has been very popular in the past, which stems from the fact that the model works exceedingly well at low bit rates. In addition, the LPC parameters preserve

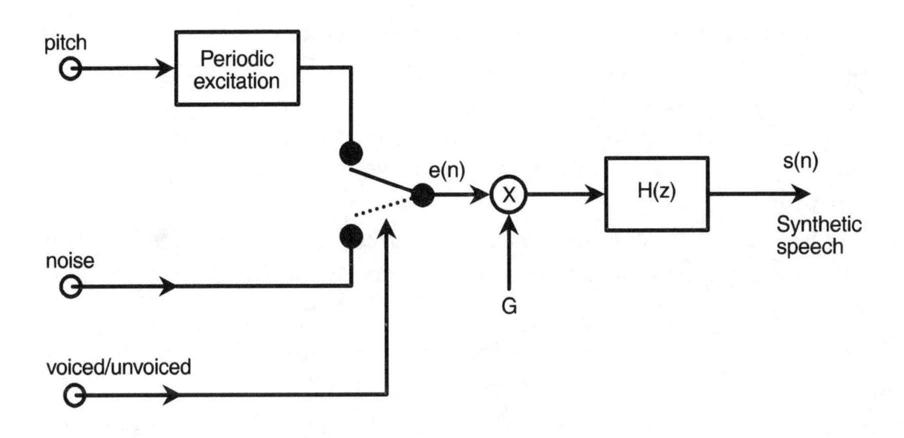


**FIGURE 6.30.** LPC vocoder

sufficient information about the speech that LPC can be used for speech recognition. For example, in a security system LPC can be used both for the speaker identification and the vocal response mechanism.

In LPC the vocal tract is modeled as an "all pole" (IIR) filter (Figure 6.30). If we include the gain $G$, we can express the filter by its transfer function, $H(z)$:

$$H(z) = \frac{s(n)}{e(n)} = \frac{G}{1 + a_1 z^{-1} + \ldots + a_p z^{-p}}$$

where $p$ is the order of the filter, $e(n)$ the excitation and $s(n)$ the synthesized speech output. Expressing this in the time domain:

$$s(n) = Ge(n) - a_1 s(n-1) - \ldots - a_p s(n-p)$$

In other words, the output speech sample is a linear combination of $p$ previous speech samples plus the excitation signal, hence the name linear predictive coding. Note that the filter coefficients $a_i$ change for each frame of speech.

In the analyzer we first window the speech input, usually with a Hamming window of between 20–40ms. The window (frame) repetition is every 10–30ms, so successive windows overlap. The choice of these two parameters depends on the final bit rate desired: The smaller the two values, the higher the bit rate and the quality of speech required.

As we mentioned earlier, voiced sounds typically have a significant low-frequency content. To ensure that the LPC vocoder accurately reproduces all frequencies, the signal is first passed through a pre-emphasis filter, which boosts the high frequencies. A first-order FIR filter is generally sufficient. The synthesizer contains a complementary de-emphasis filter.

The vocal filter coefficients are estimated by making an estimate of the speech signal $s'(n)$, which is a combination of $p$ previous samples:

$$s'(n) = -a_1 s(n-1) - \ldots - a_p s(n-p)$$

then determining the values of $a_i$ such that the mean squared error:

$$\sum_n (s(n) - s'(n))^2$$

is minimized, leading eventually to a set of equations:

$$a_1 r(0) + a_2 r(1) + \ldots + a_p r(p-1) = -r(1)$$

$$a_1 r(1) + a_2 r(0) + \ldots + a_p r(p-2) = -r(2)$$

$$\vdots$$

$$a_1 r(p-1) + a_2 r(p-2) + \ldots + a_p r(0) = -r(p)$$

where:

$$r(i) = r(-i) = \sum_{n=0}^{N-i-1} s(n) \cdot s(n+i)$$

The values $r(i)$ are called the autocorrelation coefficients. The mathematical derivation of the solution to this set of equations can be found in Papamichalis (ed.) [1987]. The interesting point to note here is that these equations are well matched to our DSP structure – they are all based on the multiplication and addition of a series of previous data values and coefficients.

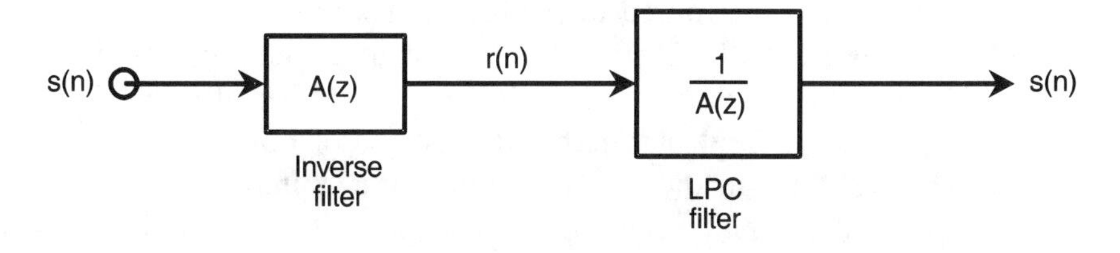


**FIGURE 6.31.** Basics of the RELP vocoder

### Residual Excited Linear Prediction (RELP) Vocoder

Looking back to Figure 6.30, we defined the synthesizer filter to be $H(z)$ where:

$$H(z) = \frac{s(n)}{e(n)} = \frac{G}{1 + a_1 z^{-1} + \ldots + a_p z^{-p}}$$
$$= \frac{1}{A(z)}$$

It has been shown that if we pass the speech signal through the inverse filter $A(z)$ we get an error signal, called the residual error. Obviously, if we pass the residual error back through the LPC filter we can reconstruct the speech signal (see Figure 6.31). The RELP vocoder efficiently codes this residual error signal $r(n)$ for transmission. RELP may be viewed as having a similar relationship to LPC as DPCM has to normal PCM.

The residual signal $r(n)$ has a relatively flat spectrum since the LPC filter coefficients were chosen such that the LPC spectrum is the envelope of the speech signal. Also, the excitation information is more or less uniform over the residual spectrum so we can get away with coding only a small range of frequencies, 0–1kHz. At the synthesizer this baseband is repeated to generate the higher frequencies and the final signal is the input to the LPC filter. The RELP vocoder is used with transmission rates of 9.6kbits/s. The advantage of RELP is that it has a higher speech quality than LPC for the same bit rate, but at the cost of a greater processing requirement.

### Multipulse Excited Linear Predictive Coding (MLPC)

Multipulse excited LPC is an extension of LPC that goes some way to overcome the mechanical sound of synthesized speech. This mechanical sound is attributed to the fact that we use only two types of excitation pulse in LPC: one for voiced and one for unvoiced speech. As the title suggests, in MLPC we use a number of pitches for each syllable independent of whether it is voiced or not.

The MLPC algorithm effectively computes the $n$ pitches in a speech signal sequentially. Once it has located one pitch it subtracts it from the signal and begins to look for the next. Obviously, the determination of the relevant pitch information is a complex task which we will not pursue here.

Nevertheless, MLPC offers better speech performance for the same bit rates as LPC. As with any system the trade-off is against the complexity of the algorithm. MLPC is used with transmission rates of 4.8kbit/s or 9.6kbit/s and can be implemented on a second-generation TMS320 device.

### Code Excited Linear Prediction (CELP) Vocoders

The CELP vocoder is a relatively new speech-encoding scheme first introduced by Atal and Schroeder [1984]. This vocoding scheme offers high-quality speech at low bit rates, but does demand a great deal of computational resource. Implementing the original systems required several hundred MIPs (million instructions per second). Much of the recent research has concentrated on reducing this load in order to make it possible to implement the algorithm on available silicon.

It is possible to implement a U.S. federal standard CELP at 4.8kbit/s on a TMS320 floating-point DSP. Nonstandard CELP algorithms at 4.8kbit/s and 8kbit/s will execute on fixed-point TMS320 DSPs. These vocoders differ slightly in their implementation, but for simplicity we shall discuss the general form of CELP and leave the intricacies for you to uncover from further reading.

CELP is based on vector quantization. Consider a second-order LPC filter, i.e.,

$$H(z) = \frac{1}{A(z)} = \frac{G}{1 + a_1 z^{-1} + a_2 z^{-2}}$$

Essentially, when we chose the parameters $a_1$ and $a_2$, we picked one point in the $p$ dimensional spectral space, where $p$ was the order of the filter, in this case two. In simple LPC and RELP we then digitize each of the coefficients separately. If we use eight quantization levels (three bits) for each, we can interpret this as dividing the $p$ spectral space into a grid as shown in Figure 6.32.

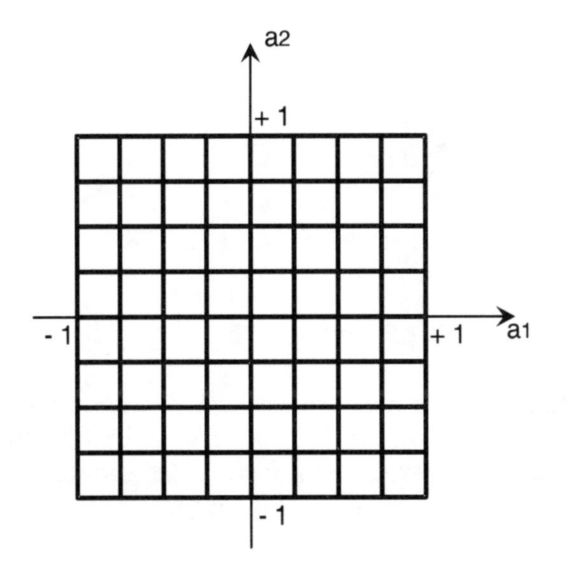


**FIGURE 6.32.** Vector spectral space showing grid of eight possible values of each of $a_1$ and $a_2$

This is efficient if we believe that we are equally likely to get speech frames in all of the available positions on the grid. In fact this is not the case: If speech is analyzed it is found that some positions are much more likely than others. Vector quantization makes use of the fact that some of the positions are redundant. The most likely positions are represented by a vector and all these vectors are stored in a codebook at both analyzer and synthesizer.

The various vectors are referenced by a distinct number or index. An incoming speech frame is then compared with the codebook and only the relevant index and its distance from the origin are transmitted. This codebook gives rise to the name Code Excited LPC.

The perceived quality of vector quantization depends on how easy it is to match sounds to this list of "standard" sounds. It is limited by the large amounts of storage necessary for the codebooks and the computational speed required during encoding if the codebook is large. CELP uses a codebook of length 512 or 256 vectors. Speech quality degrades gracefully as the codebook size decreases.

The disadvantage of most CELP coding schemes is that they fail to represent the high frequencies in speech at bit rates in the order of 6kbit/s or lower. For this reason the newer CELP schemes are actually

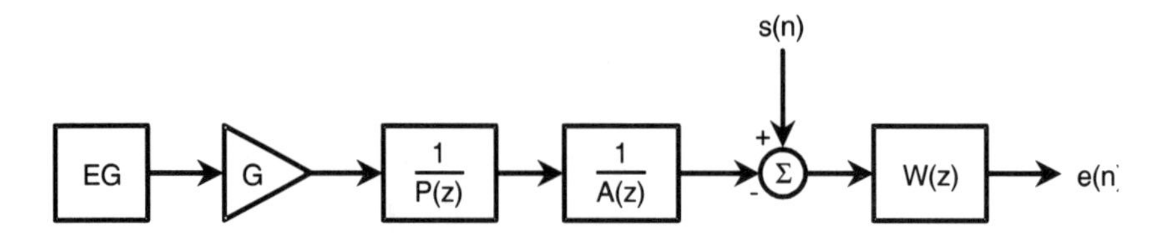


**FIGURE 6.33.** CELP encoder

a combination of CELP and MLPC using a limited number of excitation pulses.

Figure 6.33 shows a block diagram of the basic CELP encoder. The excitation generator, *EG*, produces an innovation sequence, which is multiplied by a gain factor, *G*, and fed through the pitch synthesizing filter, $\frac{1}{P(z)}$. This output is then fed into the synthesizing filter, $\frac{1}{A(z)}$. The difference between this synthesized speech and the original speech spectrum, $s(n)$, is then weighted according to a subjective error criterion, $W(z)$, to produce the error signal, $e(n)$. The coefficients of the filters, the gain $G$ and the excitation generator are varied to minimize $e(n)$, which is encoded using vector quantization.

There are many papers on the latest CELP/MLPC coders that provide a deeper insight to their workings. Several are listed in the references at the end of this chapter.

### Vector Sum Excited Linear Prediction (VSELP)

The main difference between VSELP and CELP is the method of searching the codebook. In VSELP there are actually two codebooks of 128 vectors each. As we mentioned earlier, the codebook position is the value that is actually transmitted. In VSELP this is referred to as the basis vector. Therefore, with 128 positions, each codebook is represented by seven basis vectors. The transmitted information, called the code vector, is generated as a linear combination of the basis vectors, using coefficients of 1 or −1.

The other major difference between CELP and VSELP is the favored implementation. The U.S. Department of Defense (DoD) CELP algorithm uses floating-point arithmetic, and VSELP uses a fixed-point algorithm. Although fixed-point DSPs can perform floating-point mathematics they are obviously not as efficient at performing CELP as their floating-point counterparts. The advantage of

VSELP is that we use fixed-point DSPs, which presently are more widely available and generally cheaper.

## IMAGE CODING

So far, we have spent a considerable amount of time talking about the encoding and transmission of speech signals. Today, we are also offered the ability to transmit pictures (images) either down a telephone line or over private leased lines. In the future, with the new switching and transmissions standards called SDH (Synchronous Digital Hierarchy) in Europe and the Far East and SONET (Synchronous Optical Network) in the Unites States, we shall be able to hold video conferences with friends and colleagues anywhere in the world. Videophones will become a standard feature and cable TV will be piped into the home using the same cable.

Let us first look at how our standard TV pictures are produced. This example describes the UK 625-line PAL (Phase Alternate Line) system. Three signals are used, one called luminance and two chrominance signals. The luminance signal represents the light/dark nature of the picture and the chrominance signals represent the color. The required frequency to represent all the information in the picture is high, as can be shown by the following simple analysis.

The TV screen is made up of 625 lines, each with 625 picture elements in it. If the picture is scanned once every 20ms, or 50 times per second we can just about avoid the eye perceiving a "flickering" of the picture. The number of elements in the picture is then:

$$N = 625 \times 625 = 3.9 \times 10^5$$

In the worst case we should have alternate elements of black, then white. We also include the screen aspect ratio of $\frac{4}{3}$, which affects the spatial and temporal frequency that the eye requires. The number of cycles in the picture is then:

$$N' = \frac{4}{3} \times \frac{N}{2} = 2.6 \times 10^5$$

For a frame rate of 50 per second, our information rate would be:

$$f = 2.6 \times 10^5 \times 50 = 13\text{MHz}$$

In practice only 575 lines are displayed, the remainder being used for flyback and synchronization. In addition, we must allow time for horizontal flyback and synchronization, all of which have the effect of reducing the maximum frequency to 11MHz.

From this we can see that the required bandwidth for a television picture is very high – it is reduced by interlacing alternate lines and transmitting them only once every other frame. Although this brings the bandwidth down to around 6MHz, we can easily see that video signals use a great deal more bandwidth than speech signals.

In current analog TVs, the video signal is transmitted using amplitude modulation and the audio signal is transmitted using frequency modulation. To reduce the required bandwidth, the video signal is actually coded using vestigial AM, which transmits all of one sideband and a small amount of the low frequencies of the other sideband. In this way the low frequencies, which contain the most significant information, are preserved. Even so, the total bandwidth used is approximately 8MHz.

If we wished to transmit the monochrome video signal as a digital waveform, it would require a sampling frequency of around 12MHz (twice the information rate of 6MHz). In order to avoid brightness contours around areas of constant luminance, we need to use seven bits per sample to represent the signal. If we then add on extra bits for complex coding algorithms and error correction, the result is a bit-rate of around 120Mbit/s! If we then take into account the chrominance signals, the basic bit rate would be in the order of 200Mbit/s.

As we mentioned earlier, differential PCM can often be used with video signals owing to the amount of correlation between adjacent samples. An alternative coding method is to transform the picture into a set of transform coefficients. We could use FFTs, but this results in values that have real and imaginary parts and so produces just as complex a code as the PCM signal.

Transform coding makes use of the fact that most of the information is in the low-frequency components. A square picture transforms into a square set of transform coefficients (Figure 6.34), where the amplitudes at the bottom right-hand corner are often so small that they can be ignored. The remainder are then encoded using ADPCM or DPCM, etc. Obviously, after decoding, the receiver must perform the inverse transform to reconstruct the signal.

The most common technique in image coding is to use a discrete cosine transform (DCT) since it generates coefficients that are independent. In practice the picture is split into blocks, $8 \times 8$ pixels in size,

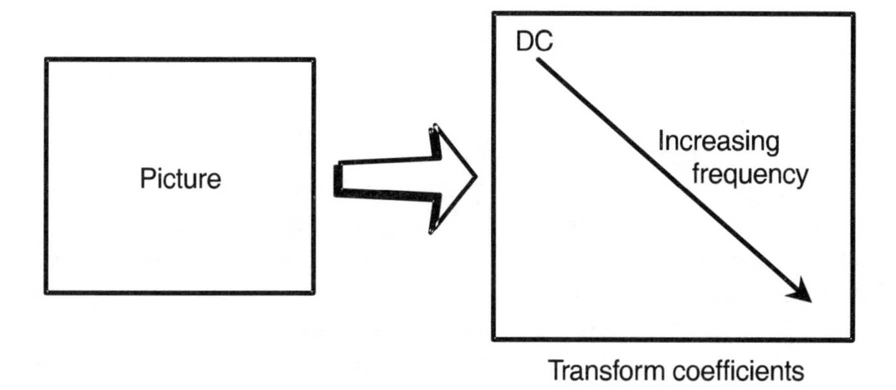


**FIGURE 6.34.** Basics of transform coding – the picture transforms into a square set of transform coefficients

which are individually transformed. High-amplitude, low-frequency coefficients are transmitted first, producing progressive picture buildup that is useful in bandwidth-limited systems. By using transforms and some form of DPCM, we can reduce the bandwidth required to transmit a full-screen image from 200Mbit/s to around 34Mbit/s.

Another popular encoding scheme is variable-length encoding or entropy coding. Entropy coding can be used either on its own or as an additional bit-rate reduction method after DPCM. The technique exploits the statistical redundancy in a signal where all the possible types of output do not occur with the same probability.

The entropy coder has to construct output words that the decoder can interpret both in length and content. This is ensured by making the start of all the output words different. Several methods have been devised to do this, the most popular by Huffman [1951]. The entropy of a signal can be used as an indication of what levels of bit-rate savings can be achieved using variable-length coding. The entropy of a signal is defined starting from the information content of a codeword, which is related in the following manner to the probability of its occurrence, $P_i$:

$$\text{information} = -\log_2 P_i$$

Therefore, the information content of a rare codeword is very big. The entropy of a signal source is then defined as the average information content per codeword:

$$\text{entropy} = H = -\sum_i P_i \log_2 P_i$$

It has been shown that this entropy value represents the minimum number of bits per symbol into which the source can be coded, assuming that the samples are independent. As we said earlier, cosine transforms generate independent coefficients. This is one of the reasons they are popular in image coding. A variable-length Huffman code gives an average word length that is close to the entropy value, implying a very efficient scheme.

A normal twisted-pair of copper wires used for domestic purposes has a bandwidth of only a few kilohertz. It doesn't take a mathematical genius to work out that this means that it is virtually impossible to transmit high-quality moving pictures using this channel. Nevertheless, it is possible to produce some low-quality videophones that work over the telephone line.

Generally, these videophones make use of the fact that the people using them are stationary, i.e., standing/sitting in one position. The amount of change in the picture from frame to frame is then proportionally small. By encoding only the difference between consecutive frames, we can considerably reduce the required bit rate. In fact, videophones use a method of encoding called motion vector estimation, which is based on this property.

At the beginning of this section we spoke briefly about the new transmission standards that will be in widespread use by the year 2000, i.e., SONET and SDH. These new standards are based on the use of very high bit rates of 155MHz, 622MHz and 2.4GHz, transmitted over coaxial cable and fiber optics. These will bring the full realization of multimedia systems where images, speech and data can be transmitted simultaneously over the same channel. The onset of high-bandwidth channels has lead to the development of new standards for image and video transmission combined with speech. In fact, there are three standards that we shall briefly look at: H.261 from the CCITT, JPEG (Joint Photographics Experts Group) for still images and MPEG (Moving Pictures Expert Group), both from the International Standards Organization (ISO).

### H.261 Video Compression

In the late 1970s the European telecommunications industry identified the need for international collaboration on the implementation of audiovisual services. The research project COST211 (Co-Operation in the field of Scientific and Technical research) and the CEPT (Conference on European Posts and Telecommunications) Working Group TR1 resulted in a European standard videoconferencing codec specification for the transmission of 625-line, 25-pictures/s PAL television at 2Mbit/s. The demand for the same services in North America also prompted the development of a compatible codec at 1.544Mbit/s for 525-line, 30-pictures/s NTSC systems. Standard conversion between PAL and NTSC was incorporated in the codec.

This system concept lead to the CCITT recommendations H.120 and H.130. In the 1980s not all codecs manufactured complied to these standards, partly because only the Europeans were involved in the systems definition. In the United States and Japan proprietary systems arose that were adopted by corporate users, who then were unable to interface to the standard codecs.

The SGXV/1 Specialist Group was set up to develop a worldwide standard at 384kbit/s and its multiples. The H.261 standard developed an algorithm for $p \times 64$kbit/s, which covers bit rates from 64kbit/s up to 2Mbit/s. The lowest frequency can be used for videophones running on narrowband ISDN (Integrated Services Digital Network).

The H.261 scheme uses a hybrid DPCM/DCT technique with motion compensation. Figure 6.35 shows a simplified diagram of the scheme. The luminance signal is first sampled at 6.75MHz and coded with eight bits, The chrominance signals are then sampled at 3.375MHz. The difference between the present frame and the previous frame is then split into $8 \times 8$ blocks on which DCTs are performed. The coefficients are then coded using a variable-length Huffman coding algorithm to reduce the amount of data. An inverse DCT is performed on the quantized coefficients and the previous frame is re-added. This results in a picture very similar to the original which can be stored for subsequent use in the next frame.

The motion vector compensator works on the principle that an object that moves from one frame to the next can simply be represented by the displacement and direction in which it moves, i.e., the vector, and no information about the object needs coding. Obviously, this is an over simplification since the object may also change in some way during the translation.

The motion compensator takes each $8 \times 8$ pixel block and searches the previous image by moving the block $\pm 15$ samples in the horizontal and vertical directions in an attempt to find the best match. The

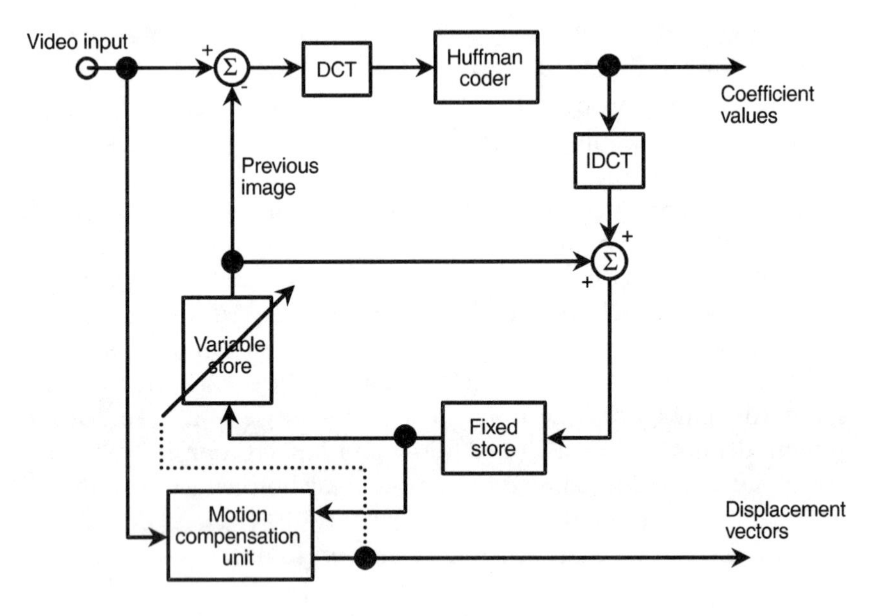


**FIGURE 6.35.** H.261 encoder

variable store shown in Figure 6.35 is adjusted to use the best approximation for the subtraction from the present frame and the resultant vector is passed on to a coder for transmission. Figure 6.36 outlines a simplified diagram of the H.261 decoder.

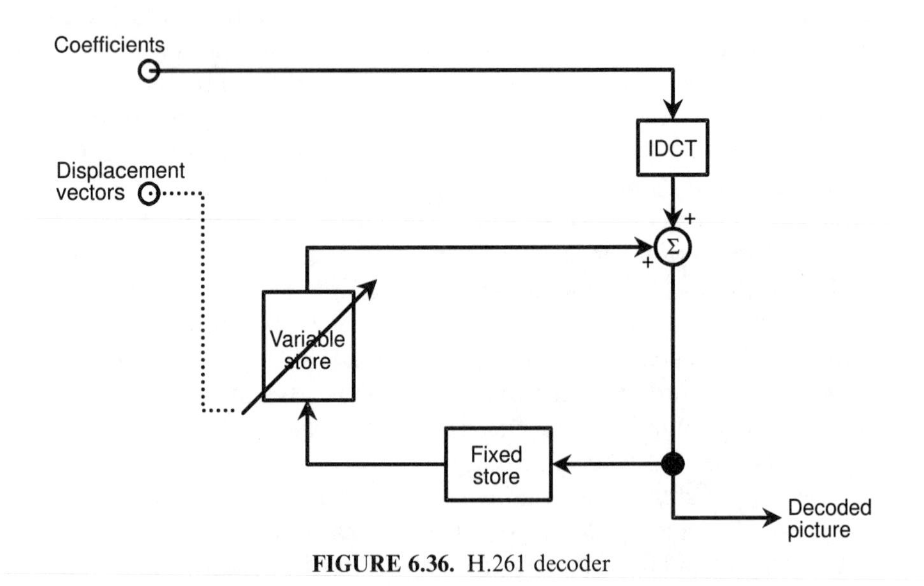


**FIGURE 6.36.** H.261 decoder

H.261 leaves manufacturers the ability to design codecs for various applications, e.g., videoconferencing, videophones, etc., at different bit rates. A videophone will probably use only a small screen and therefore the number of pictures per second required, the number of bits/pixel, the number of pixels and the coarseness of the motion compensation can be adjusted to give a much lower bit rate than for a videoconferencing system, which would use a large screen and require a higher-quality image.

There is equipment available that conforms to this standard – for example, British Telecom has an H.261 videoconferencing system. The algorithm is computationally intensive and hence most applications use a large amount of processing power. Videophones generally use floating-point or multiple fixed-point DSPs to perform the DCT and DPCM coding for low bit-rate systems (64kbit/s). A detailed study of the British telecom system, using a floating-point DSP, can be found in Kenyon and Nightingale [1992].

## JPEG

The Joint Photographics Expert Group (JPEG) proposed standard is aimed at still picture compression. It is a transform-based coding algorithm that is applicable to any type of composite color system, e.g., the standard television luminance and chrominance signals, or a system that is based on the primary colors of red, green and blue. Each color component is transformed by DCTs on $8 \times 8$ blocks and the resulting coefficients are quantized using a system dependent on the component and the frequency (Figure 6.37).

The frequency dependence allows high-frequency components to be encoded with a smaller number of bits than the more important lower-frequency information. JPEG also allows the designer to assign different quantization schemes to the different components, ensuring the most important has the highest number of bits – similar in effect to the H.261 case where the chrominance signals were sampled with only half the frequency of the luminance signal.

The coefficients are reordered into a single stream in a zigzag fashion (Figure 6.38) and the whole stream is Huffman coded to reduce the amount of data to be transmitted. The dc term is differentially encoded with the previous frame before Huffman coding in order to reduce the difference between this value and the subsequent higher-frequency coefficients.

The JPEG coder is simpler than the H.261 system although they

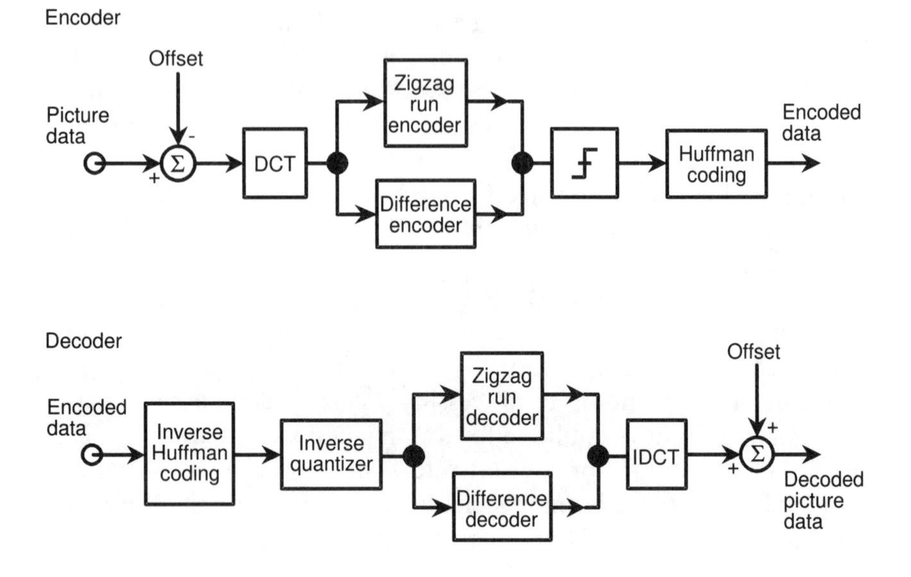


**FIGURE 6.37.** JPEG still-image encoder and decoder

contain many similar elements. On the other hand, the JPEG decoder is more complicated. In general, the main difference between the two is the use of differential encoding and motion compensation in the H.261. The JPEG system simply resets after each transmission. This is not unreasonable when we think of the aim of this standard – still-picture transmission. It does not expect another frame.

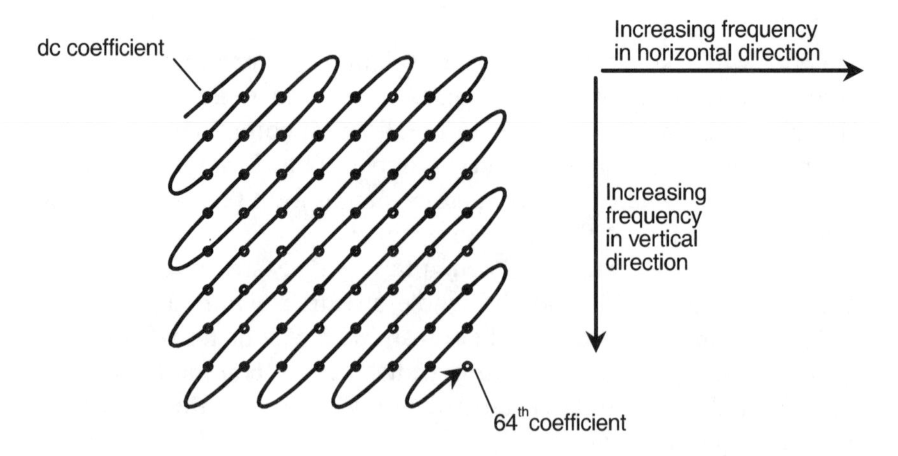


**FIGURE 6.38.** Zig-zag coding reorders coefficients into a single stream

Investigations are under way to use JPEG to compress moving pictures. There is still uncertainty as to the best way to achieve optimum results.

### MPEG

The Moving Pictures Expert Group (MPEG) proposed standard (MPEG 1) is aimed at full-motion compression on digital storage media. It is similar to H.261 in that it uses interframe and intraframe techniques to reduce the bit rate. As it is aimed at digital storage media such as CD-ROM, it cannot exceed their present limit of 1.5 Mbit/s.

MPEG uses the same type of intraframe compression as JPEG. Taking into account the interframe compression, MPEG is more than twice as effective at compression than "moving" JPEG. On the down side, as it is more complex than JPEG, it needs significantly more computing power.

Once the picture has been split into blocks and the DCT performed on each of the blocks, the coefficients are coded with either forward or backward predictive coding or a combination of both. After this, the blocks are uniformly quantized using a matrix of quantization steps chosen to achieve the desired output bit rate.

Although JPEG and MPEG were devised to solve two different tasks, JPEG also has some advantages in video compression. First of all, it is a symmetrical technique, so that the same method is used to decompress the information as to compress it. In this way, we can build a single compression processor to perform both tasks. In some video teleconferencing systems JPEG is being used for this reason.

### Future image-coding techniques

We have looked only at how the standards for image compression have evolved. At the time of this writing, the standards are still developing. There are many debates as to which standard is best and whether JPEG or MPEG is better at full-motion video encoding. To ensure the latest information, it will be necessary to scan the electronic journals. Nevertheless, we hope we have demystified some of the terminology that is always found in electronics and given some insight into the commonly used DSP techniques.

When reading about the future of video techniques you will often see references to fractals and wavelets. These are a mathematician's delight and because of this we have not included anything about them in this book.

## REFERENCES

When researching information for this chapter we found that all the books on speech coding were very mathematical and it was necessary to sift the math for the more interesting information on how the systems work. Below, you will find some books that are better than most, for example Papamichalis' book [1987], which is aimed at practical implementations and includes examples of algorithms coded for TI DSPs.

Unfortunately, the newer image-coding schemes are too new to find many textbooks on, so we have been limited to listing some papers. These do have the advantage that you can find some with more text than mathematics!

Atal, B.S. and Schroeder, M.R. [1984]. "Stochastic Coding of Speech at Very Low Bit Rates," *Proceedings of ICC 1984,* pp. 1610–1613.

Bateman, A. and Yates, W. [1988]. *Digital Signal Processing Design,* Pitman Publishing, London, UK.

Bonomi, M. [1991]. "Multimedia and CD ROM: An Overview of MPEG and JPEG," *CD ROM Professional,* November 1991, pp. 38–40.

Carlson, A.B. [1981]. *Communication Systems,* Second Edition. McGraw Hill, New York.

CCITT [1984]. *32kbit/s Adaptive Differential Pulse Code Modulation (ADPCM),* CCITT Recommendation G.721.

Charbonnier, A., Maitre, X. and Petit, J.P. [1985]. "A DSP Implementation of the CCITT 32kbit/s ADPCM Algorithm," *Proceedings of IEEE International Conference on Communications*, Vol 3, pp. 1197–1201.

Clark, A.P. [1983]. *Principles of Digital Data Transmission,* Pentech Press, Devon, UK.

Esteban, D. and Galand, C. [1977]. "Application of Quadrature Mirror Filters to Split Band Voice Coding Schemes," *Proceedings of the 1977 IEEE International Conference on Acoustics, Speech and Signal Processing,* Hartford, CT, pp. 191–195.

Gonzales, C.A. and Viscito, E. [1991]. "Motion Video Adaptive Quantization in the Transform Domain," *IEEE Transactions on Circuits and Systems for Video Technology,* Vol 1, No 4, December 1991, pp. 374–378.

Habibi, A. [1971]. "Comparison of Nth Order DPCM Encoder with Linear Transformations and Block Quantization Techniques," *IEEE Transactions on Communications,* December 1971, pp. 948–956.

Huffman, D.A. [1951]. "A Method For The Construction of Minimum

Redundancy Codes," *Proceedings of the IRE,* No. 40, pp. 1098–1101.

Jayant, N.S. and Noll, P. [1984]. *Digital Coding of Waveforms,* Prentice Hall, Englewood Cliffs, NJ.

Kenyon, N. and Nightingale, C. [1992]. *Audiovisual Telecommunications,* Chapman & Hall.

Kingsbury, N.G. and Amos, W.A. [1980]. "A Robust Channel Vocoder for Adverse Environments," *Proceedings of the 1980 IEEE International Conference on Acoustics, Speech and Signal Processing,* April 1980, pp. 53–60.

Kondoz, A.M.; Lee, K.Y. and Evans, B.G. [1989]. "Improved Quality CELP Base Band Coding of Speech at Low Bit Rates," *Proceedings of the 1989 IEEE International Conference on Acoustics, Speech and Signal Processing.*

Lin, Kun-Shan (ed.) [1987]. *Digital Signal Processing with the TMS320 Family, Volume 1,* Prentice Hall, Englewood Cliffs, NJ.

Martin, J.D. [1991]. *Signals and Processes, A Foundation Course,* Pitman Publishing, London.

Papamichalis, P.E. (ed.) [1987]. *Practical Approaches to Speech Coding,* Prentice Hall, Englewood Cliffs, NJ.

Sandbank, C.P. [1990]. *Digital Television,* John Wiley and Sons, NY.

Schulthei, M. and Lacroix, A. [1989]. "On the Performance of CELP Algorithms for Low Rate Speech Coding," *Proceedings of the 1989 IEEE International Conference on Acoustics, Speech and Signal Processing.*

# 7

# Practical DSP Hardware Design Issues

This chapter starts by summarizing the basic hardware alternatives available for DSP system design. We then look in detail at some of the issues and considerations involved in implementing a system using a single-chip DSP device. This is the chapter that contains most of the practical advice on actually putting your algorithms into practice on some DSP hardware. You will also find supporting information in Chapter 8, which describes a typical development flow for a DSP design in the context of design decisions.

## HARDWARE ALTERNATIVES FOR DSP

It is possible to implement DSP algorithms on any computer hardware, for example a PC, but the rate at which you want to process information determines the optimum hardware platform for your application. There is a basic divide here between real-time and non-real-time DSP applications. Real-time applications include audio playback processing, speech compression in mobile communications, etc. Non-real-time applications may be seismic data processing, data compression for storage on prerecorded CD-ROMs, and so on. This book is only concerned with real-time digital signal processing, so PCs, etc. are ruled out of our discussion.

There are four categories of hardware platform widely used for real-time DSP implementation: general-purpose DSP devices, special-purpose DSP devices, bit-slice processors and general-purpose microprocessors. Let's look at each one briefly and see where it fits into the picture.

General-purpose digital signal processors include the TI TMS320 family, the Motorola DSP56000 and DSP96000 families, the Analog Devices ADSP2100 and 21000 families and the AT&T DSP16 and DSP32 families. These devices are fully programmable and therefore very flexible. They have fast cycle times and faster variants are released at regular intervals as technology and design techniques progress. Their performance and flexibility allow the selection of a "company standard" DSP, where one hardware platform can be used for different DSP tasks. This brings savings in staff training, development tools, software and so on. Common devices are produced in huge volumes for many customers, bringing down unit costs. These general-purpose devices will have a variety of on-chip peripherals and general-purpose instruction sets. They can often be used as the only processor in an embedded system, not needing the support of a general-purpose host processor. This makes them extremely cost effective in a very broad range of applications.

A number of companies such as Zoran, Plessey and SGS Thomson Microelectronics (STM) produce special-purpose DSP devices targeted at specific tasks. Examples may be an FIR filter chip, an FFT device or a convolution engine. These are useful where a well-defined basic DSP operation needs to be done in the minimum amount of time. Special-purpose DSP devices are able to perform their limited functions much faster than general-purpose DSPs because of their dedicated architecture. The weaknesses of this approach are the lack of standard development tools, the consequent long design cycle, the cost of the components and the likely need for a host processor to be designed into the system.

Bit-slice components, such as the AMD29000 family, offer a more general-purpose approach to high-speed DSP design. These components are basically building blocks including multipliers, ALUs, sequencers, etc., that are joined together to build a custom DSP architecture. Performance can be significantly higher than with general-purpose DSP devices. The disadvantages are similar to those for the special-purpose DSP devices: lack of standard microprocessor design tools, extended design cycles and high component cost. Bit-slice processors are most widely used in military DSP applications where cost and design cycle time are secondary requirements to performance.

Any computer can be used for digital signal processing and this applies to common microprocessors such as the Motorola 68000 or Intel i86 families. These devices do not have the architecture or on-chip facilities required for efficient DSP. In particular they lack a hardware multiplier. A floating-point multiply operation, which will take a single 50ns clock cycle on a typical DSP device, may take up to 44 cycles of 35ns on a 68030, a factor of 31 slower. The only time general-purpose processors are likely to be used for DSP is where a small amount of signal processing is required in a much larger system. Their real-time DSP performance does not compare well with even the cheapest general-purpose DSP devices, and they would not be a cost-effective solution for any serious DSP task.

## USING A GENERAL-PURPOSE DSP DEVICE

The remainder of this chapter concentrates on the implementation issues involving general-purpose DSP devices, such as members of Texas Instruments' TMS320 family. Let's first look at the architecture of two of the most common devices, the TMS320C25 and the TMS320C30. We shall contrast and compare these two devices and use them throughout the remainder of this chapter to illustrate key points.

A block diagram of the TMS320C25 is shown in Figure 7.1. The device is a 16-bit fixed-point processor with a Harvard architecture, i.e., it has separate program and data memory spaces (see Chapter 1). On-chip resources include a mask-programmable program ROM, multiplier, ALU, shifters and three blocks of internal RAM. One of these may be configured either as program or data RAM. Off-chip accesses to memory are via the multiplexed data and address buses. These buses are used for any access to the off-chip data or program memory or I/O spaces. This implies that off-chip program and data cannot be accessed simultaneously.

The TMS320C30 is a quite dissimilar device. Its structure is shown in Figure 7.2. The most obvious difference is that it does not have a Harvard architecture. Program, data and I/O are all mapped into the same 16M word address space. Parallel access to different areas of memory is provided by the multiple on-chip buses and the dual off-chip buses. The device is a 32-bit floating-point processor. Main functional blocks include two blocks of RAM, an instruction cache, the multiplier, ALU, registers, a DMA controller, mask-programmable ROM, serial ports and timers.

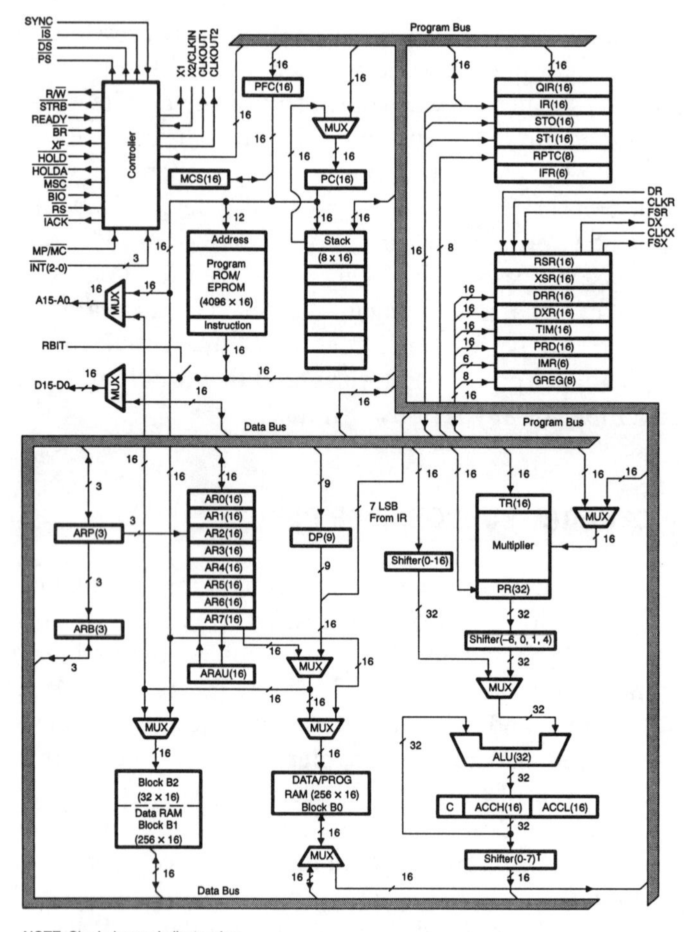


FIGURE 7.1. Block diagram of the TMS320C25

We shall not consider the detail of these two devices except where necessary to explain some of the points we are making. Detailed technical information is contained in the user's guide for each device. The devices we have chosen are widely used and broadly representative of other fixed- and floating-point devices.

## FIXED- AND FLOATING-POINT DSP DEVICES

When browsing through manufacturer's data it soon becomes clear that the most basic division between processors is that between

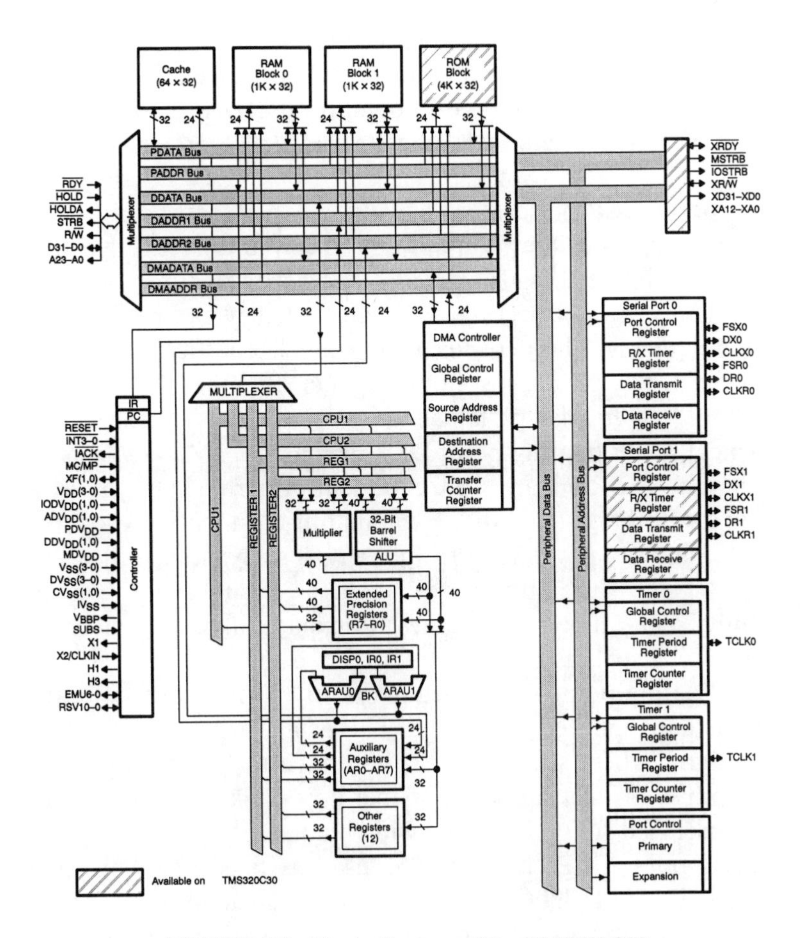


**FIGURE 7.2.** Block diagram of the TMS320C30

fixed-point and floating-point devices. Fixed-point processors are either 16-bit or 24-bit devices; floating-point processors are usually 32-bit devices.

A typical 16-bit fixed-point processor, such as the TMS320C25, stores numbers in a 16-bit two's complement integer format, which we shall describe shortly. The important aspect is that although coefficient and data values are only stored with 16-bit precision externally, intermediate values are kept at 32-bit precision within the internal accumulator, so cumulative rounding errors during calculations are not normally a problem. It is possible to use a pseudo-floating-point format in a fixed-point DSP, but the speed penalty is so great that it would only be applicable for system management tasks. DSP functions usually need the greater speed of the true fixed-point format. This also means that the use of a high-level language compiler with a fixed-point DSP is something of a compromise. We shall look more closely at the issue of high-level language (HLL) compilers in the next chapter.

Fixed-point DSP devices are usually cheaper than their floating-point counterparts as they contain less silicon and have less external pins. Fixed-point devices also have faster clock cycle times, of as little as 25ns. This does not necessarily mean that they can perform DSP tasks more quickly. This depends on how much a particular device can do in one clock cycle, not just how long a clock cycle is. We shall say more about "specmanship" later.

A typical 32-bit floating-point DSP such as the TMS320C30 stores a 24-bit mantissa and an 8-bit exponent. Intermediate values are stored in a 40-bit register with a 32-bit mantissa and 8-bit exponent. A 32-bit floating-point number format gives a dynamic range from plus $(2-2^{23}) \times 2^{127}$ to minus $2^{128}$. Remember, though, that resolution is still only 24 bits at best and quantization may still be a consideration in applications such as professional digital audio. The vast dynamic range of floating-point DSPs means that dynamic range limitations may be virtually ignored in a design. This is in sharp contrast to fixed-point designs, where scale factors often have to be applied and the designer must protect against possible accumulator overflow. This point is discussed fully in the following section.

Floating-point DSPs may also be used as fixed-point or integer processors. The TMS320C30 has a full set of instructions to allow it to operate as a 24-bit fixed-point device with the capability of storing 32-bit results.

Floating-point processors are generally more expensive than fixed-point devices, mainly because they use more silicon area, but also because they tend to use larger and more expensive device packages – twin 32-bit buses use a lot of external pins! Floating-point devices may be needed in applications where either gain coefficients are varying in time (e.g., adaptive systems), signals and coefficients have a large dynamic range, or where large memory structures are required, such as in image processing. Other cases where floating-point devices can be justified are where development costs are high, and production volumes limited. The faster development cycle for a floating-point device may easily outweigh the extra cost of the components. Floating-point processors also allow the efficient use of high-level language compilers and reduce the need to fully identify the system's dynamic range.

We shall now look in detail at the differences between the fixed- and floating-point number formats and the implications these have for DSP system design.

## FIXED-POINT ARITHMETIC

In binary format, a number can be represented as signed magnitude, where the left-most bit represents the sign and the remaining bits represent the magnitude:

+52 (decimal) = 34 (hex) and is represented as 0011 0100 (binary)
−52 (decimal) = −34 (hex) and is represented as 1011 0100 (binary)

We have already mentioned that fixed-point DSP devices use a number format called two's complement. A positive number is

| | MSB | | | LSB | | |
|---|---|---|---|---|---|---|
| Bit value | $-2^3$ | $2^2$ | $2^1$ | $2^0$ | | |
| | -8 | 4 | 2 | 1 | | |
| | 0 | 1 | 0 | 1 | = 4+1 | = 5 |
| | 1 | 1 | 0 | 1 | = -8+4+1 | = -3 |
| Least positive | 0 | 0 | 0 | 1 | | = 1 |
| Most positive | 0 | 1 | 1 | 1 | = 4+2+1 | = 7 |
| Least negative | 1 | 1 | 1 | 1 | = -8+4+2+1 | = -1 |
| Most negative | 1 | 0 | 0 | 0 | | = -8 |

FIGURE 7.3. Example of 4-bit two's complement number formation

represented as a simple binary value. The differences from conventional binary arithmetic come with negative numbers. Figure 7.3 shows an example of positive and negative two's complement numbers, using 4-bit numbers. This principle easily extends to 16-bit numbers.

The great advantage of two's complement arithmetic is that only one adder is required for both positive and negative numbers. An addition will always give the correct result for both addition and subtraction (adding a positive and negative number). An added benefit is that, if the final result is known to be within the processor's number range, an intermediate overflow can be ignored as the correct final result will still be produced. With four bits, remembering the most significant bit is the sign bit, the largest positive number that can be represented is +7, and the largest negative number −8. By simple extension from Figure 7.3, a 16-bit number can vary between +32,767 and −32,768.

It is worth noting that two's complement numbers are a programmer's convention and not usually provided by the analog-to-digital conversion hardware. Negative numbers have to be converted from signed magnitude into two's complement. Figure 7.4 shows the conversion process.

For those of you not familiar with two's complement binary arithmetic, refer first to Figure 7.5 for a description of addition and subtraction. Note particularly that in the addition of four numbers,

**The differences from conventional (signed magnitude) binary arithmetic come with negative numbers. To get -56 (decimal) as a two's complement number, we have to perform a little sum:**

| | | |
|---|---|---|
| **$-56_{10}$ represented as signed magnitude binary** | | **10111000** |
| **Strip off sign bit** | | **00111000** |
| **Invert all bits to get one's complement** | | **11000111** |
| **Add one to get two's complement** | **+** | **1** |
| **Two's complement is** | | **11001000** |

**Therefore, -56 (decimal) is represented as 11001000 in two's complement binary. Adding the two's complement representations of 56 and -56 gives:**

| | | |
|---|---|---|
| | | **00111000** |
| | **+** | **11001000** |
| **as we would expect** | | **00000000** |

FIGURE 7.4. Forming negative two's complement numbers

| Addition | | Subtraction | |
|---|---|---|---|
| | 00 0101 ( 5) | | 01 1100 (+28) |
| | 00 0100 ( 4) | | 11 1100 (- 4) |
| | 00 1101 (13) | | 11 0000 (-16) |
| | 00 0111 ( 7) | | 11 1110 (- 2) |
| carry 1 | 1 | no carry | |
| carry 1 | 1 | no carry | |
| carry 10 | 1 0 | carry 1 | 1 |
| | | carry 10 | 10 |
| | 01 1101 (29) | ignore carry | 10 |
| | | | 00 0110 (+6) |

FIGURE 7.5. Addition and subtraction with two's complement numbers

the carry is sometimes not just to the adjacent column, but may be to two columns to the left. For example, where five ones are added the result is 101, carrying two columns. A subtraction in two's complement arithmetic is the same as the addition of a positive and negative number. The first step is to negate (find the two's complement of) the number to be subtracted and then perform an addition using this negative number. The two's complement is formed as shown in Figure 7.4. We need to understand what to do with the addition of the most significant bits. In this case the addition produces a carry to the left of a one. In fact we can ignore this carry, as the negative number effectively has a one in each bit location to the left of its own most significant bit (111010 is the same as 11010, and so on).

If you're struggling at this point, get out a pen and paper and try a few examples using 4-bit numbers. Refer to Figure 7.3 to ensure you are forming the two's complement numbers correctly. There is no

| | | |
|---|---|---|
| Multiplicand | 0100 | (+4) |
| Multiplier | x1101 | (-3) |
| | 0100 | |
| | 0000 | |
| | 0100 | |
| Remembering that the MSB is a sign bit, the final partial product is the two's complement of the multiplicand | 1100 | |
| | 1110100 | (-12) |
| In an 8-bit accumulator, this will be | 11110100 | (-12) |

FIGURE 7.6. Multiplication with two's complement numbers

substitute for trying this arithmetic out for yourself. That's how everyone finally gets the hang of two's complement. The add and subtract instructions of a fixed-point DSP will do all of the work for you, but it's important to understand what's happening so you can fully debug the code.

For a more complicated example look at Figure 7.6 for a description of a multiplication. The steps are exactly the same as for decimal multiplication except for the final partial product when the multiplier is negative. As the most significant bit is the sign bit, the final partial product is the two's complement of the multiplicand. This is produced as shown in Figure 7.4 by inverting all the bits of the multiplicand and adding one. The example of multiplying two negative numbers is a little more complicated. Note that the first two partial products will be negative numbers and must have their sign bit extended to the left to preserve their correct value in the final addition. The final partial product is again the two's complement of the multiplicand.

The two's complement binary representation described above does not have any binary point and does not represent fractions. In signal processing, it is common to represent numbers as fractions. The reason for this is that when multiplying fractions (e.g., $0.99 \times 0.99$), the result will always be less than one and there is never an overflow. Where dynamic range is so severely constrained this is an obvious advantage. We need to use an implied binary point to represent binary fractions, and this is where things can get tricky. Hopefully, the following short explanation will make you confident in tackling code that uses binary fractions.

The location of the binary point affects neither the arithmetic unit nor the multiplier in the DSP. It affects only the accuracy of the result and the location from which the result will be read. The binary point is purely a programmer's convention and has no relation to the hardware at all.

In order to make the most of the 16-bit architecture of a TMS320 fixed-point device, numbers are usually represented in what is known as Q15 format. The number following the letter Q represents the quantity of fractional bits. This implies that, in Q15 format, each number is represented by 1 sign bit, 15 fractional bits and no integer bits. Similarly, a 13-bit number in Q12 format will have 1 sign bit, 12 fractional bits and no integer bits. Q15 is the common fixed-point format used by the TMS320C25. 13-bit Q12 numbers are used by the Multiply Immediate (MPYK) instruction, where the 13-bit constant is a part of the instruction word.

| | MSB | | | LSB | | |
|---|---|---|---|---|---|---|
| Bit value | $-2^0$ | $2^{-1}$ | $2^{-2}$ | $2^{-3}$ | | |
| | -1 | 0.5 | 0.25 | 0.125 | | |
| | 0 | 1 | 0 | 1 | = 0.5+0.125 | = 0.625 |
| | 1 | 1 | 0 | 1 | = -1+0.5+0.125 | = -0.375 |
| Least positive | 0 | 0 | 0 | 1 | | = 0.125 |
| Most positive | 0 | 1 | 1 | 1 | = 0.5+0.25+0.125 | = 0.875 |
| Least negative | 1 | 1 | 1 | 1 | = -1+0.5+0.25+0.125 | = -0.125 |
| Most negative | 1 | 0 | 0 | 0 | | = -1 |

**FIGURE 7.7.** Example of 4-bit Q3 binary fractions

The principles of forming Q format numbers are easily understood with reference to Figure 7.7. This shows some 4-bit Q3 numbers, i.e., there is 1 sign bit, 3 fractional bits and no integer bits.

The principles of addition and subtraction are exactly the same as for binary integers and are shown briefly in Figure 7.8. Note that the binary point remains static exactly as in decimal addition and subtraction. An example of multiplication is shown in Figure 7.9. Once again this uses two 4-bit Q3 numbers. The result is in Q6 format.

**Addition (Q5)**

| | |
|---|---|
| | 0.0 0101 (0.15625) |
| | 0.0 0100 (0.125) |
| | 0.0 1101 (0.40625) |
| | 0.0 0111 (0.21875) |
| carry 1 | . 1 |
| carry 1 | . 1 |
| carry 10 | .1 0 |
| | 0.1 1101 (0.90625) |

**Subtraction (Q5)** (adding positive and negative numbers)

| | |
|---|---|
| | 0.1 1100 (+0.875) |
| | 1.1 1100 (-0.125) |
| | 1.1 0000 (-0.5) |
| | 1.1 1110 (-0.0625) |
| no carry | . |
| no carry | . |
| carry 1 | . 1 |
| carry 10 | 1.0 |
| ignore carry | 10. |
| | 0.0 0110 (+0.1875) |

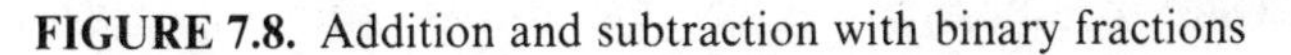

**FIGURE 7.8.** Addition and subtraction with binary fractions

| | Binary | Decimal |
|---|---|---|
| Multiplicand (Q3) | 0.100 | +0.500 |
| Multiplier (Q3) | 1.101 | -0.375 |
| | 0100 | 2500 |
| | 0000 | 3500 |
| | 0100 | 1500 |
| | 1 100 | |
| Answer in Q6 | 1.110100 | -0.187500 |

**FIGURE 7.9.** Multiplication with binary fractions

Compare each step with the decimal example alongside and the shift in location of the binary point will become obvious. We have not found a comprehensive yet simple treatment of the subject in any textbook, but once you have grasped the basic idea it should become fairly clear.

Referring back to the TMS320C25, the most common form of multiplication is with two Q15 numbers. The result is in Q30 format and is also a fraction. The result has 30 fractional bits, 2 sign bits and no integer bit.

| | |
|---|---|
| $-0.5$ | 1.100 0000 0000 0000 |
| $\times +0.5$ | 0.100 0000 0000 0000 |
| $-0.25$ | 11.11 0000 0000 0000 0000 0000 0000 0000 |

To store the result as a Q15 number, a left shift of one bit is performed to eliminate the extra sign bit and the left-most 16 bits are stored. Our result above is therefore stored as 1.110 0000 0000 0000, the 16-bit two's complement representation of –0.25.

A multiply can never give an overflow with fractions, but successive additions may. The TMS320C25, in common with other TMS320 fixed-point devices, has a mechanism to protect against overflow and indicate if it occurs. This is the *saturation mode*, where the 32-bit accumulator fills to its maximum or minimum value but does not roll-over. This has a similar effect to the "clipping" of an analog wave-form. There is a status register bit which is set if saturation occurs.

The saturation mode is very useful, as unpredictable results would occur with accumulator roll-over. This could be disastrous in a control system. Just consider the effect of a serious arithmetic error resulting from accumulator roll-over in a temperature controller for a reflow solder process. Having said this, the clipping effect is also generally undesirable in DSP systems. Scaling factors should be applied to inputs or intermediate results to ensure that it does not happen. Scaling is discussed briefly in Chapter 3 and in the following section.

## QUANTIZATION EFFECTS AND SCALING IN FIXED-POINT DSPS

In applications such as control and high-quality digital audio, the effect of finite word lengths can be critical. In a fixed-point processor, signals and coefficients must be scaled to fit in the dynamic range and word length of the processor. Like quantization effects in an analog-to-digital converter, finite word length effects appear as noise in the system.

When designing a digital filter, the filter coefficients are quantized to be less than or equal to the word length of the processor so that they can be stored in the program memory. This could mean that the performance of the filter will be slightly different from its design specification. The quantization becomes more significant the tighter the specification of the filter, affecting IIR filters more than FIR filters. FIR filters are unconditionally stable, but coefficient quantization can cause serious problems in IIR filters where poles may be pushed from just inside to just outside the unit circle, causing instability. In most DSP systems, coefficients quantized to 12–16 bits do not cause changes in performance. Parks and Burrus [1987] and DeFatta et al. [1988] contain detailed discussions of the effects of finite register length in digital filters.

We dealt with signal quantization caused by analog-to-digital conversion in Chapter 3. A second form of signal quantization appears when the results of signal processing are truncated or rounded. The input and output of a system may be 16-bit values, but to maintain accuracy, intermediate calculations will require greater precision. For example, a $16 \times 16$-bit multiply will require 32 bits to store the full precision of the result. As TMS320 fixed-point DSPs have a 32 bit accumulator there isn't often a problem. Of more concern is the fact that memory locations are 16-bit. Special instructions are provided to store all 32 bits of the accumulator in memory, so that they may be restored for subsequent use. Fortunately, we are normally only concerned with (at most) the 16 most significant bits of a result for conversion back to an analog signal. As long as the program is written carefully it is quite possible to ensure that rounding or truncation occurs only at the final stage of calculation.

With the use of proper scaling of signal values, coefficients or both, we can minimize or eliminate the effect of signal quantization caused by truncation, rounding or overflow. This is not the place for a detailed discussion on the selection of correct scaling factors. Discussions of the subject can be found in various texts, with Bateman and Yates [1988] giving a readable introduction and Jackson [1989] giving a particularly thorough treatment. We shall confine ourselves to a few simple guidelines.

The scaling factor should be chosen carefully, taking into account the translation of input and output values (i.e., the voltage range of the A/D and D/A converters) and the need to prevent intermediate overflow or saturation. It is possible to reduce the amount of scaling required by altering the structure of a transfer function. Cascaded forms are usually preferable to direct forms. These allow scaling to be applied to each substage, reducing signals by only as much as is necessary to prevent overflow within that stage. This is preferable to applying all the scaling required at the input as it reduces the signal quantization effect that would occur. There are some cases such as certain speech processing algorithms where cascaded forms do not give this advantage. We can use software simulators to determine the best form of scaling for our system. These are explained in more detail in Chapter 8.

Figure 7.10 shows the multiplier, ALU, accumulator and associated hardware within the TMS320C25. This includes mechanisms for dealing with the consequences of fixed-point arithmetic. The multiplier takes inputs in the form of two 16-bit or one 16-bit and one 13-bit number (multiply immediate). The shifter shown on the output of the multiplier allows the extra sign bits produced to be removed and a simple 16-bit two's complement number to be stored or output. Note also that both the lower and higher 16 bits of the accumulator may be stored separately in memory for subsequent reloading of the full 32-bit precision.

There is some extra programming effort required in using fixed-point DSPs in comparison with floating-point devices, but there are major advantages also. The lowest-cost fixed-point devices are around one-fifth the cost of the cheapest floating-point devices. If a fixed-point DSP will do the job, it is almost always the most cost-effective solution. The extra design effort will usually be insignificant compared with the component cost savings in volume production. Of course, if

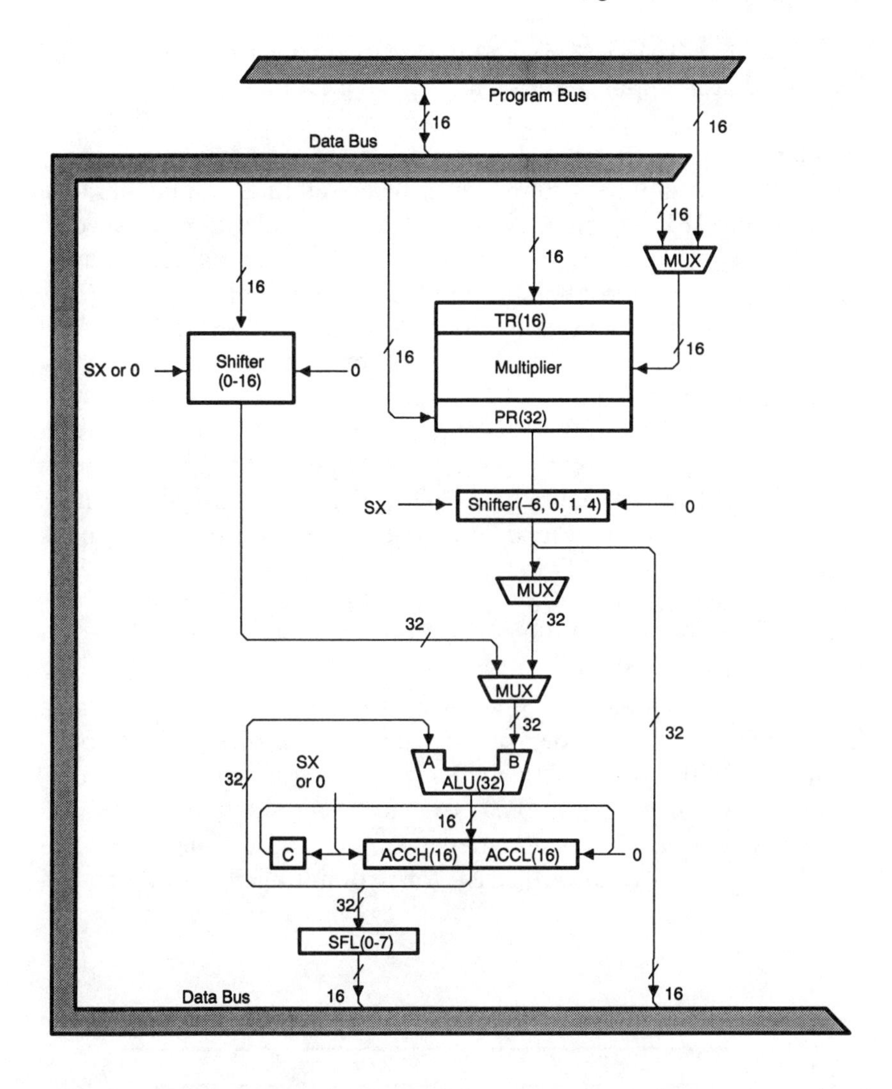


**FIGURE 7.10.** The TMS320C25 multiplier, ALU, accumulator and associated hardware

only a few systems are being made, the balance could tip in favor of the shorter design cycle with the floating-point device.

## FLOATING-POINT ARITHMETIC

The industry standard floating-point number format is defined by IEEE Standard 754.1985. Several floating-point DSP devices, including TI's TMS320C3x and TMS320C4x devices, use a special form of floating-point representation for internal calculations. This means that they are not directly compatible with the IEEE 754 standard. This is not usually a problem as compatibility is only required when inputting numbers from or outputting numbers to a device using IEEE format. For all internal calculations, no specific format is required. There are two ways to achieve compatibility: use a software conversion routine (a single instruction in the TMS320C40) or use an external conversion ASIC (Application Specific Integrated Circuit). In either case the overhead for conversion is normally insignificant when compared to overall algorithm execution time.

Figure 7.11 shows the IEEE Standard 754.1985 single-precision floating-point format. Note that this is a 32-bit format and fits within the bus width of a floating-point DSP. The standard also defines single-extended, double and double-extended precision formats. Figure 7.12 shows the TMS320 floating-point number format. The most obvious difference is in the location of the sign bit, but there are also some more detailed differences. For example, the exponent is a two's complement number.

The reason for the difference from the IEEE standard is that a hardware multiplier is easier to design and smaller (therefore cheaper) in the TMS320 format. With single-cycle hardware conversion, compatibility is not an issue, but as device technology improves and geometries shrink the size advantage will become less important and new DSPs will move toward IEEE format multipliers. For a detailed

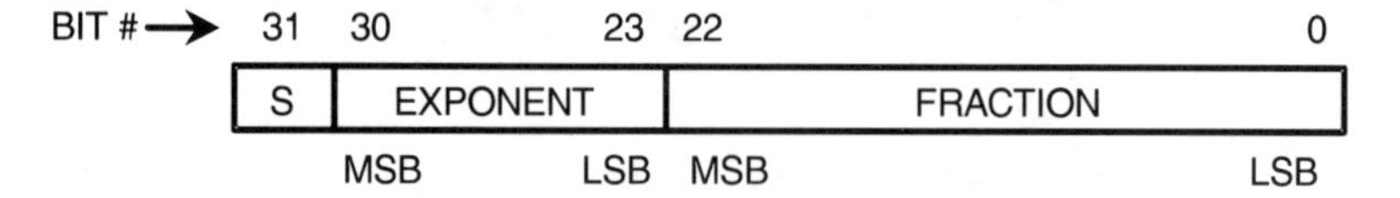


S is the sign bit of the mantissa (0=positive, 1=negative)

EXPONENT is an unsigned 8-bit field that determined the location of the binary point of the number being encoded

FRACTION is a 23-bit field containing the fractional part of the mantissa

LSB is the least significant bit of a field

MSB is the most significant bit of a field

**FIGURE 7.11.** IEEE 754 floating-point number format

| 31 | 24 | 23 | 22 | 0 |
|---|---|---|---|---|
| EXPONENT | | S | FRACTION | |
| MSB | LSB | | MSB | LSB |

The decimal value (v) of some number X is determined as follows:

$v = \{(-2)^S + (.FRAC)\}2^{EXP}$

Where: S is either 0 or 1

FRAC is the decimal equivalent of FRACTION
EXP is the decimal equivalent of EXPONENT

An alternate way of describing the TMS320 format mantissa is as follows:

$\overline{S}$S.FRACTION - note that the bit to the left of the binary point is implied and is the complement of the sign bit.

**FIGURE 7.12.** TMS320 floating-point number format

description of the IEEE 754 single-precision and the TMS320 floating-point formats, refer to Papamichalis (ed.) [1991]. This provides a full list of special cases, such as zero and ± infinity, for both formats.

Both floating-point formats provide a dynamic range roughly from $2^{128}$ to $2^{-128}$ with a resolution of 24 bits. When using a floating-point processor, factors such as scaling and dynamic range are not normally an issue, though it is worth remembering that the 24-bit resolution means that we cannot necessarily discount signal quantization effects, particularly in applications such as professional audio.

## DSP SYSTEM SPEED CONSIDERATIONS

We have all seen the advertisements for PCs: 16MHz 386SX, 33MHz 386, 50MHz 486, and so on, but when we read comparative reviews of similarly specified machines, suddenly we find that some machines are much faster than others in real performance. Poorly designed systems can be handicapped by slow screens, slow disks, poor memory architecture, etc.

Specifications have become the way to sell PCs and there is also a lot of "specmanship" in the semiconductor industry. Digital signal processors do not escape from this. Figures are eagerly quoted for MFLOPS, MIPS, MOPS, MBPS, etc. These figures are, of course, always accurate, but they don't necessarily tell us a great deal about what a particular DSP device will do in our application. Refer to

**TABLE 7.1. Floating-point processor mnemonics**

| | |
|---|---|
| MFLOPS | Million Floating-Point Operations Per Second - The number of floating-point multiply, add, subtract, store, etc. operations that the processor can perform in a second. Care is needd to ensure that quoted figures are sustainable, i.e. that they can be maintained for the whole of the execution period of a piece of code. Some quoted figures are for peak performance only. |
| MOPS | Million Operations Per Second - The total number of operations the processor can perform in a second. This will include DMA accesses, data transfers, I/O operations etc. It gives a rough guide to the total processing and I/O capabilities of the processor |
| MIPS | Million Instructions Per Second - The number of machine code instructions the processor can execute in a second. The key issue is how much can each instruction do. The MFLOPS figure is usually a more reliable measur of processor performance. |
| MBPS | Mega-bits per second - Usually refers to the bandwidth of a particular bus or other I/O port. Provides a measure of data throughput. |

Table 7.1 for an explanation of the mnemonics associated with floating-point DSPs.

The figures place each processor into a broad performance category, but are those MFLOPS a peak or a sustained value? What happens to device performance if we need to access a lot of data off-chip? The answers to these questions are often far from clear. We may need to put in a lot of effort to predict our system performance before committing to a choice of DSP device. The following paragraphs give a few hints as to what to look for.

## Accessing Memory Resources

Digital signal processing systems have complex requirements for memory. Program code and data values used in fast DSP routines need to be stored in fast memory to achieve rapid access. Initialization code and coefficients can usually be stored in slower and cheaper memory. There may also be a need to store some data in EEPROM, or other nonvolatile memory. An example is the sets of room characteristics used in a programmable audio sound-field processor.

Most modern DSP devices have a straightforward memory interface and only simple logic is required where address decoding is necessary. Most DSP memory interfaces are intended for interfacing to static RAM or EPROM, as interfacing to dynamic RAM is quite clumsy. This is because static RAM is faster and more suited to the time-critical applications DSP systems are designed for. Some floating-point DSPs with large memory address spaces, such as the TMS320C40, have interface signals to allow relatively fast page-mode or static-column-decode access to dynamic memories. This keeps memory cost to realistic levels in large systems without a drastic speed penalty.

You will find that almost all published specifications for DSP devices relate to operations that are performed entirely on-chip. This means that the specification assumes no accesses off-chip are required. This is often unrealistic as we may have large off-chip data and coefficient tables or rapid flow-through of data from an external analog-to-digital converter (ADC). Look for the amount of RAM available on a device. Can it store your time-critical program instructions and all the data and coefficients needed? If not, there is likely to be a speed penalty for the off-chip accesses. Look to see what this is. Typically one wait-state may be required for each word accessed off-chip. In some cases it may be more.

In early DSPs there was a requirement for the address decoder to be connected to some wait-state logic, which would insert the different numbers of wait-states for whichever physical block of memory was being accessed. The circuitry is not complex, but any external logic has a penalty in system cost and memory timings. Devices such as the TMS320C30 and TMS320C5x have software-programmable wait-states whereby different blocks of the address range will have wait-states inserted by the processor when they are accessed. All the designer needs to do is to connect his memory (or peripherals) to the appropriate address block and load the wait-state register. These processors also power-up with the maximum number of wait-states, so an external slow boot EPROM may be used.

To understand some of the implications of memory usage and interfacing, let's look at an example of an FIR filter algorithm on the TMS320C25. Figure 7.13 shows the on-chip RAM resources of the device. We can see that there are three blocks, adding up to 544 words. One of these blocks (B0) may be configured as either program or data memory. The core of an FIR in the TMS320C25 is an MACD instruction (**M**ultiply and **AC**cumulate with **D**ata move). The MACD is usually preceded by a repeat (RPT) instruction and executes repeatedly from the instruction register (effectively a single-location instruction cache) until the FIR loop is complete. The repeat counter has a maximum value of 255, so for longer loops the RPT/

MACD sequence must be repeated. The MACD multiplies a coefficient in program memory with a data sample in data memory, stores the result in the product ($P$) register, adds the previous value of the product register to the accumulator and moves the data sample by one memory location. See Chapter 4 for a full explanation of MACD.

If our filter has 256 or less taps both coefficients and data can be stored entirely in the internal RAM blocks. In this case the FIR filter core takes $3 + n$ cycles, where $n$ is the number of taps. If our filter has more than 256 taps, but less than 544, the data samples may be stored

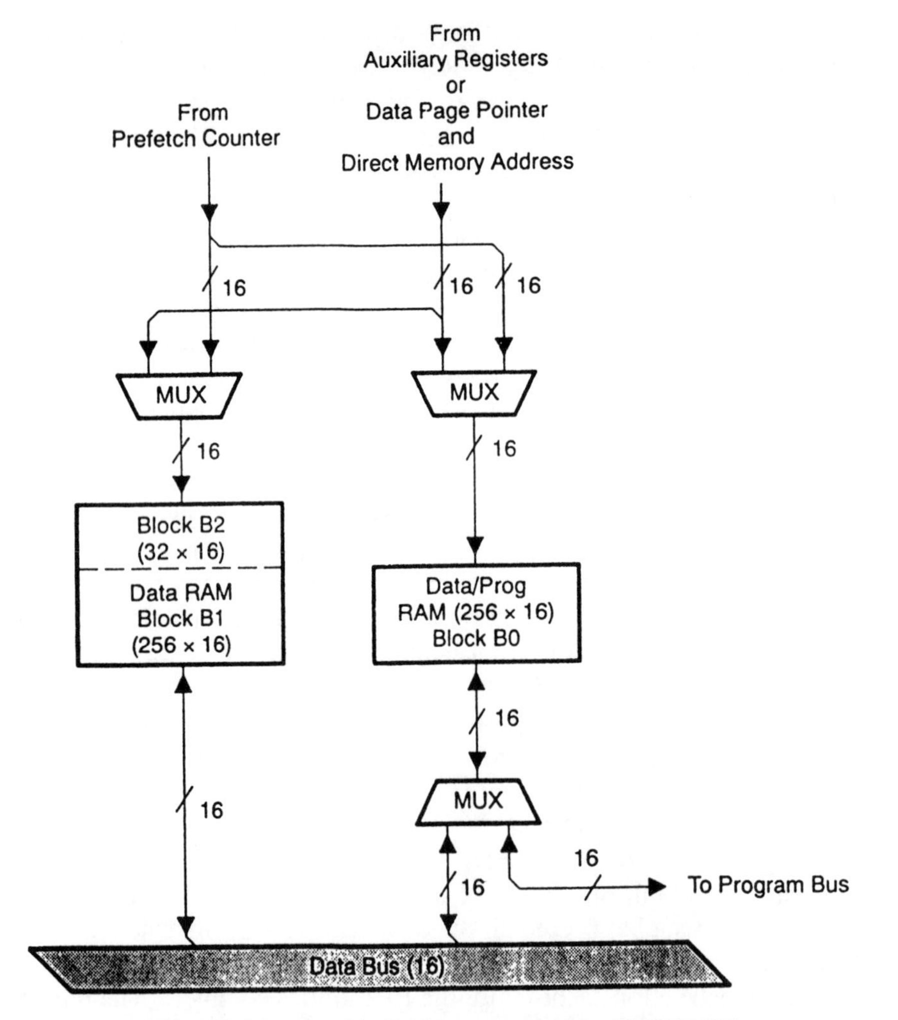


**FIGURE 7.13.** On-chip RAM resources of the TMS320C25

in internal RAM, but the coefficients must be stored externally. In this case the filter core takes $3 + n + np$ cycles, where $p$ is the number of wait-states required to access off-chip program memory. Assuming the memory access time is less than around 40ns, the TMS320C25 can access external program memory with zero wait-states. In this case there is no speed penalty for the off-chip access. If the external program memory requires one wait-state we can see that the time taken to calculate the FIR doubles.

Finally, if our filter has more than 544 taps, both the data samples (at least some of them) and the coefficients must be held in external memory. The TMS320C25 has only one memory bus for external accesses to data, program and I/O spaces, so accesses to data samples and the coefficients must be interleaved. This causes the filter to take $3 + 2n + nd + np$ cycles, where $d$ is the number of wait-states required to access off-chip data memory.

This example shows several features that we should look for: Can the program, data and coefficients all reside on-chip? Are there any restrictions on simultaneous access of on-chip resources, and if so, can the code be written to avoid these? Is there an instruction cache? What are the penalties for off-chip accesses? The answer to each of these questions will depend on the DSP device being evaluated. The operation and trade-offs will be different for each device.

The TMS320C30 takes a quite different approach to handling

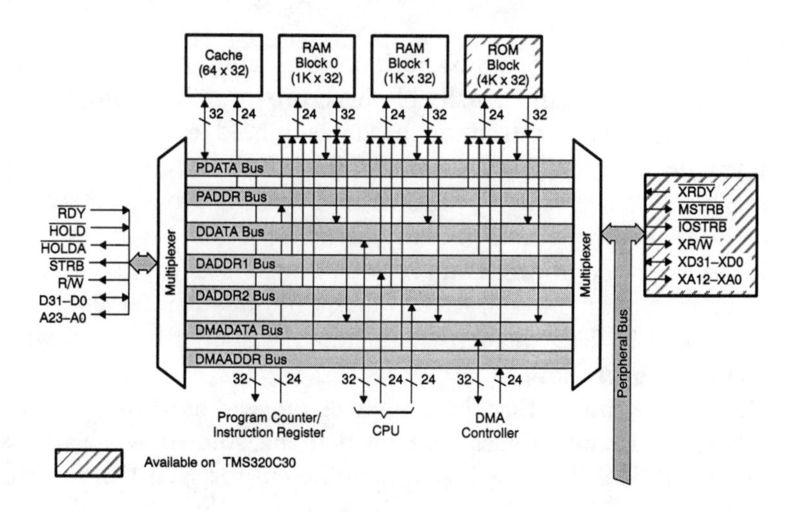


**FIGURE 7.14.** Memory structure of the TMS320C30

memory resources. We have already seen from Figure 7.2 that it does not have a Harvard architecture. The memory structure of the device is shown in Figure 7.14. Note that internal memory is split into three main blocks, ROM and two RAM blocks. Note also the internal program cache, the multiple internal buses and the dual external buses. The Harvard architecture of the TMS320C25 achieves speed through separating program and data and having the ability to access both simultaneously. The TMS320C30 achieves its speed by combining program and data into one memory space, but having the ability to access different areas of its memory at the same time.

The program, data and DMA buses are totally independent. It is quite possible for the program bus to read an instruction from the program cache at the same time as the data bus reads a value from RAM block 0 and the DMA bus writes a value into RAM block 1. The parallelism goes even further because there are two data address buses and the on chip RAM blocks are dual access. Two values can be read from a single RAM block (or one from each) in a single cycle.

In the case of a simple FIR filter there will be no problem with sizes of up to 1024 taps as data and coefficients may be held in internal RAM. For larger filter sizes up to 2048 taps, either the data or the coefficients will require external accesses. This will not slow down execution as the TMS320C30 has zero wait-state access on the off-chip primary bus. For algorithms where any two of program, data and coefficients are stored externally there may be a speed penalty as the expansion bus has access to only a limited amount of memory without wait-states.

The TMS320C30 contains both an instruction cache and a DMA controller, which can be valuable in improving algorithm execution performance. The instruction cache will hold 64 instructions for immediate access. This can be useful where external memory is slow, but it is necessary for a small loop of code to execute quickly. The DMA controller provides a separate processing unit for I/O and data transfer operations. Data may be transferred, for example, from a serial port to an internal RAM block without any intervention from the main CPU. The transfer will operate transparently, i.e., it will not affect CPU operation.

By now we can see that there is a wide range of hardware design and system partitioning issues that need to be studied while assessing candidate DSPs. These will be quite different for each DSP device, as the TMS320C25 and TMS320C30 examples illustrate. All of these issues can be assessed by careful study of manufacturer's data, but this can be very time consuming. For a significant system design it is worth getting software simulators for candidate devices and trying a few things out. This will give a feel for the real device performance in a real application.

An easier but less rigorous alternative is to look at some independent benchmarks. These will cover the standard DSP functions such as different length filters and FFTs, maybe a DCT or more complex functions. Journals such as *EDN* are a good source. Manufacturers also provide accurate benchmarks for their own devices, but beware of their comparative studies. A manufacturer only publishes comparative benchmarks for algorithms that their device is particularly good at. This may be a particular length of FFT that just fits in their device, but needs external accesses on a competitor's device. You may find that the relative performances are completely reversed for a different length of FFT.

Optimizing memory use is one of the most difficult areas for the system designer. There are so many possibilities: internal vs. external memory, looped vs. straight-line code, instruction caches, DMA controllers, etc. The objective is to perform the DSP task in the time available with the most efficient solution – getting there can be a challenge. We do not pretend to have any magic answers as each DSP device has its own strengths and weaknesses and its own characteristics when it comes to memory subsystem design. Probably the best approach is to keep an open mind when assessing possible solutions and to talk to the manufacturer's own technical specialists if you are in any doubt.

## Integration of Peripheral Devices

The level of peripheral integration within a DSP device can affect performance, system cost and design complexity. Intuitively it would seem that the more of the final system is integrated onto one piece of silicon the better. This will be true for most high-volume applications, but not always so for lower-volume general-purpose applications. There are some peripherals, such as serial ports and timers, that are found on the majority of DSPs. There are others, such as ADCs and DACs, PWM outputs, event managers, etc., that are rarer. All silicon area costs money. Timers and serial ports are sufficiently small and commonly used that the cost increase is small and the penalty for buying a chip with a timer and not using it is marginal. The situation with ADCs and DACs is quite different. It is not that easy to integrate high-precision analog circuitry with high-speed digital circuitry on one piece of silicon. There will be interference, particularly with digitally generated noise appearing on the analog ground, supply and references. Using mainly digital converters such as sigma delta will help and integrated solutions are becoming available.

## REFERENCES

There are very few general references in the area of DSP system design. Bateman and Yates [1988] makes a bold effort and contains much useful information. Perhaps the most useful approach is to look to a general DSP training course as provided by a number of different training organizations.

There are several texts dealing with the issues of using fixed-point DSP devices. Finite register length effects are rigorously covered in DeFatta, Lucas and Hodgkiss [1988] and in Jackson [1989].

For those wanting to use C, there is an excellent book by Embree and Kimble [1991] that contains a brief introduction to DSP algorithms and follows this up with a detailed description of how to implement these in C. It rather skips over the fact that most practical systems will be a mixture of C and assembler, but it is an essential book for anyone using C in a DSP application. Example source code in Kernighan and Ritchie C is provided on a disk.

There are also some independent books on designing with specific DSPs. The most substantial of these is Chassaing and Horning [1990], which deals with the TMS320C25 in great depth. This book was developed from a college lecture course and is a little mathematical in places. Despite this, it contains a wealth of practical information on using a fixed-point DSP device.

U.S. electronics journals are a good source for independent benchmarks, especially *EDN* (Electronics Design News). The September 29, 1988 issue contained a series of DSP benchmark tests. These are occasionally updated with new devices.

Manufacturers also publish extensive application volumes in support of their DSP devices. These often contain information regarding efficient algorithms for standard DSP functions and can also be a valuable source of instruction regarding DSP techniques, number formats, etc. Our discussion on fixed-point number formats is partly based on Ahmed [1991] and our discussion on floating-point numbers on Papamichalis (ed.) [1991].

Ahmed, Irfan (ed.) [1991]. *Digital Control Applications with the TMS320 Family,* Texas Instruments, Dallas, TX.

Bateman, Andrew and Yates, Warren [1988]. *Digital Signal Processing Design,* Pitman, London, UK.

Chassaing, Rulph and Horning, Darrell W. [1990]. *Digital Signal Processing with the TMS320C25,* John Wiley, New York.

DeFatta, David J., Lucas, Joseph G. and Hodgkiss, William S. [1988]. *Digital Signal processing: A System Design Approach,* John Wiley, New York.

Embree, Paul M. and Kimble, Bruce [1991]. *C Language Algorithms for Digital Signal Processing,* Prentice-Hall, Englewood Cliffs, NJ.

Jackson, Leland B. [1989]. *Digital Filters and Signal Processing,* Kluwer Academic Publishers, Norwell, MA.

Papamichalis, Panos (ed.) [1991]. *Digital Signal Processing Applications with the TMS320 Family, Volume 3,* Prentice-Hall, Englewood Cliffs, NJ.

Parks, T.W. and Burrus, C.S. [1987]. *Digital Filter Design,* John Wiley, New York.

# 8

# DSP System Design Flow

This book is aimed at engineers with little or no previous experience of digital signal processing. Therefore, in this chapter we hope to explain the DSP design process and answer some important questions about method, tools and technique. Figure 8.1 shows a generalized and abstracted DSP system design flow that everyone inevitably goes through. We will look at the components of this flow and the tools available to assist in each of the stages.

## DEFINE SYSTEM REQUIREMENTS

This chapter is concerned with the practical aspects of undertaking a design project using a single-chip digital signal processor. We do not cover in detail the more theoretical aspects of DSP system specification such as system requirements definition, signal analysis, resource analysis and configuration analysis. These are important considerations in any design, but the detail required to introduce these topics is beyond the scope of our book. There is an excellent discussion of the subject, with a full worked example, in DeFatta et al. [1988]. This is thoroughly recommended for anyone embarking upon a major DSP system design. Our analysis of the design process begins after the basic requirements of the system are determined.

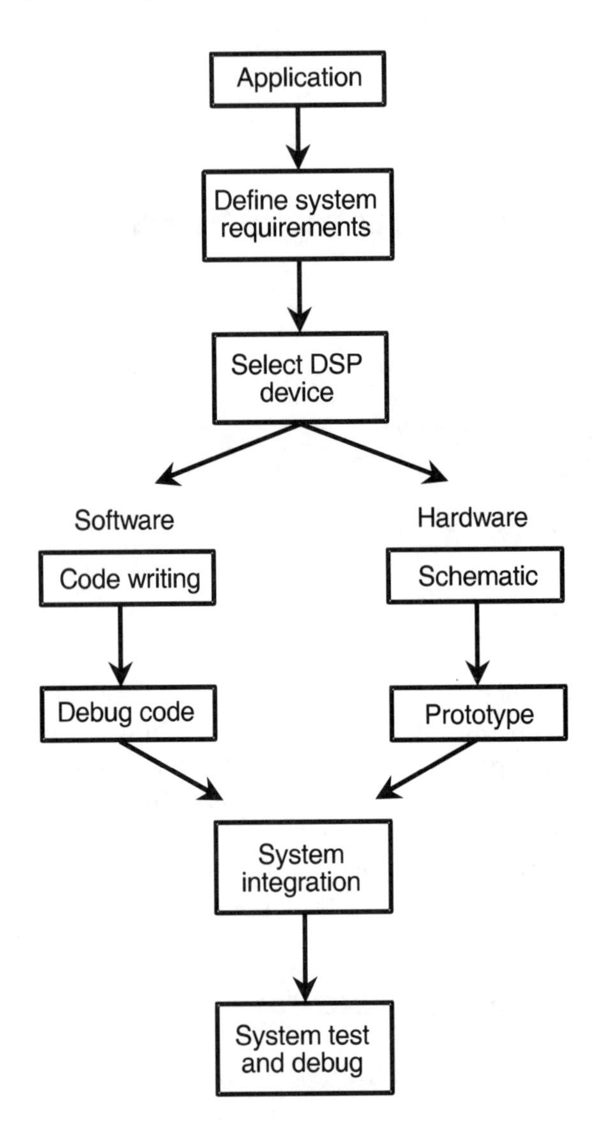


**FIGURE 8.1.** Simplified DSP design flow

## SELECTING A DSP DEVICE

It will be necessary to have a full understanding of the processing requirements of the system you are to design before a choice can be made regarding which of the available DSP devices to use. Many devices can be eliminated from consideration at an early stage due to lack of power, insufficient resolution, cost and so on. This will probably still leave a number of candidate devices for which further analysis will be required.

We discussed some of the hardware considerations that may be important in the previous chapter. It will be necessary to obtain full data on each device and do a detailed analysis of how your algorithm and interface requirements will match to each device. This is also the time to consider the way in which you will implement your design. The processor you finally choose will need to have a full set of development tools available. Basic requirements are:

Detailed design documentation
Code development tools at assembler or higher level
Tools for testing design functionality
Application notes or other design assistance

Optional requirements are:

High-level language compiler for modular and maintainable software (C, ADA, FORTRAN, etc.)
Application libraries
Manufacturer's training workshop
Real-time operating system
Low-cost tools for assessing device suitability
Ability to test full system in real time and debug hardware
Reduced time to market through enhanced support
Low-cost prototyping
Workstation-based development environment

The objective is to select the device that will allow us to meet the project timescales and implement the most cost-effective solution. For high-volume applications this will probably mean that the cheapest device that can do the job will be chosen. For low- to medium-volume applications there will be a trade-off between development tool cost and the effectiveness and cost of the device. For very low-volume applications it will often make sense to use a device that is easy to design in or one with the lowest-cost development tools. DSP vendors are now very conscious that a low cost-of-entry route to DSP projects is required for a number of applications. Texas Instruments offers low-cost evaluation modules and designer's kits for its most popular devices.

In the PC add-on environment it may be best to look for one of the many standard boards available from vendors such as Atlanta Signal Processors or Loughborough Sound Images. This approach reduces the development process to one of software design only, possibly benefiting from the use of an operating system, C-compiler and libraries of standard routines supplied with the board.

There is a chance that the choice of the DSP device for your design will be an iterative process. In other words, you may not make the correct choice the first time. It could be that unforeseen problems appear when you try benchmarking some code, or even that you find that a cheaper and less powerful device could be used. Commonly, the design specification will alter and force a re-think in the chosen implementation. We can guard against the first two cases by being thorough in the research of suitable devices. It may even be worth buying some lower-cost development tools such as designer's kits or software simulators for several devices and exercising the code before committing to a single device.

## APPLICATION DESIGN

There are parallel strands within the design of any DSP application. The most obvious are hardware and software design. In large systems these will be the responsibility of different individuals or teams of individuals. In smaller designs the hardware and software may be more interdependent. It might be best for one engineer to be responsible for both. Even if this is not the case, it is desirable for the hardware and software engineers to have a good understanding of each other's craft. It should be fairly clear by now that there is a lot of interdependence between hardware and software in DSP, so engineers can no longer operate in isolation. The ideal DSP designer will be a true "system" engineer, capable of understanding issues with both hardware and software.

By the time the full design process starts it will be necessary for all engineers working on the project to have a full understanding of the DSP device chosen—its strengths, weaknesses, and unique characteristics. While this knowledge can undoubtedly be obtained through practice, it may be far more efficient to take advantage of the device-specific training courses offered by the DSP vendors or other bodies. These will cost a thousand dollars or more for an in-depth workshop,

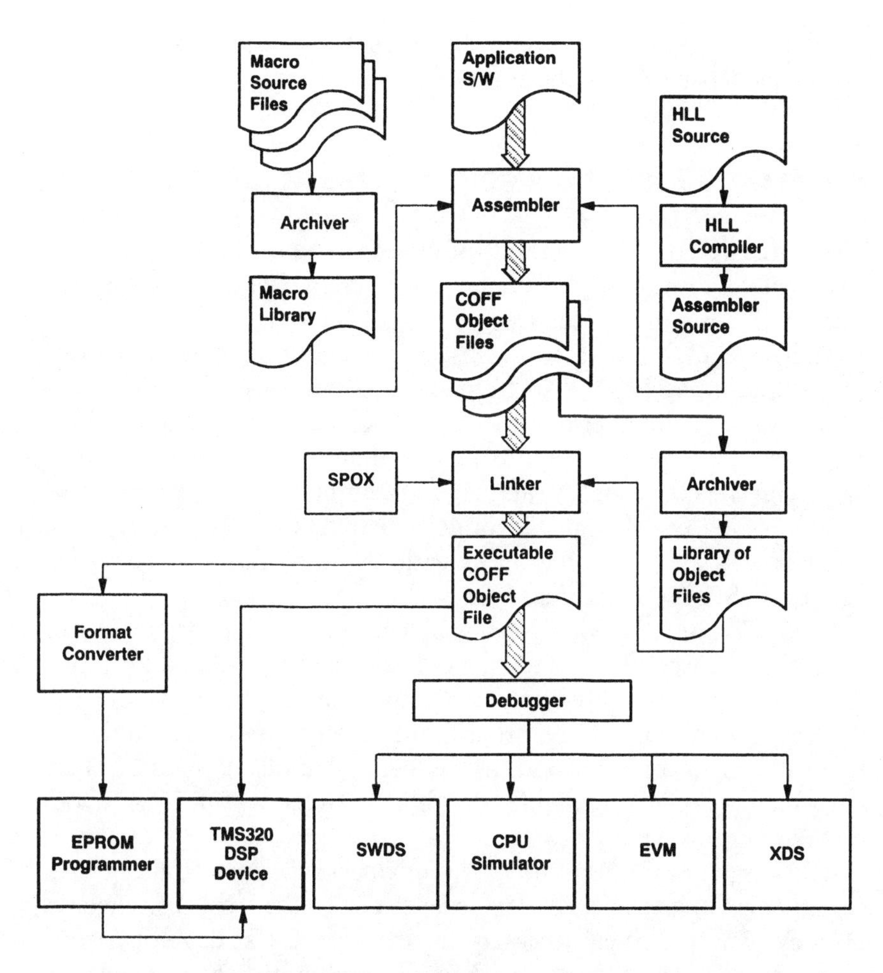


**FIGURE 8.2.** TMS320 family development tools

but can save several weeks of design effort. This reduction in time-to-market could make a project; increased time-to-market could break it.

## SOFTWARE DESIGN

Figure 8.2 outlines the various software development tools supporting the TMS320 family and how they fit together. At the bottom of the diagram are the various platforms that can be used for testing and debugging the code. The software development process is fairly logical with either assembler or C, or a combination of both, being used to generate the system code. This can then be tested on a variety of platforms depending on the stage of development.

## GENERATING ASSEMBLER SOURCE – THE ATTRACTIONS OF C

There are two common methods of writing code for DSP devices: use an assembler or use a high-level language. Often the ideal solution is to work with a mixture of the two.

There are a lot of attractions in using a high-level language (HLL) for programming DSPs. Most obviously, a language like C is understood by a large number of engineers, which means that the code may be understood by many more people than if it were written in a particular DSP's assembly language. Assembly languages tend to be mnemonic based, though algebraic assemblers are becoming more popular. Also, with very few modifications, C programs can be ported between DSPs from different manufacturers.

Generally, high-level languages offer a structured environment for code writing. Functions, data structures, typed variables and all the rest make for reasonably easily understood and maintained code, but there is still an onus on the programmer to ensure the "readability". By far the most common HLL in use for DSP applications is C. There are other compilers available, notably ADA and FORTRAN, but these are not as widely accepted.

The first objection that any DSP engineer will have to using a C-compiler is the speed penalty. In order to design the most cost-effective DSP system, it is often necessary to squeeze the last drop of performance from a particular device. This implies that the code will need to be very "tight" or carefully crafted, to make the most of the available performance.

Modern C-compilers are very efficient, but they will never be a substitute for hand-crafted assembler code. The C-compilers available for the TMS320 processor family are among the best around, but for a typical LPC speech coding algorithm there is still a factor of 1.5 speed penalty compared with code written in assembly language. In some cases this would be perfectly acceptable and the advantages of C will win out, but there will be many systems that cannot withstand this penalty. These will typically be ones where the sampling rate is high or a great deal of calculation needs to be squeezed in between samples.

In these cases the solution is simply to mix C and assembly code. All DSP C-compilers generate an intermediate assembly language stage and time-critical routines can be edited at this point. Alternatively, an assembly routine may be either called as a function or in-line coded into the C program. A library of hand-optimized functions may be built up and brought into the code where required.

With the improvement in C-compiler technology and the sophisticated debug and development tools available, a mixture of C and assembly language is now the most popular way of writing code for a large DSP system. It combines the advantages of the convenience and portability of C with the raw speed of assembly language. In typical DSP software, the proportion of hand-optimized code that will be necessary may only be around five percent of the total. This five percent of the source code will also be where the processor will spend the majority of its time, the core of the DSP algorithm.

Even taking into account the simplicity of software design when using a C-compiler it is worth bearing two points in mind. First, there is a wide range of quality of C compilers available for DSP devices, some of which are far less satisfactory than others. Second, it is always necessary to use an assembler and linker within the HLL code development process (see Figure 8.2).

## TESTING THE CODE

There are two basic methods for trying out your DSP code: a software simulator or some form of hardware platform. Software simulators run on a host, such as a PC or Sun workstation, and mimic the behavior of a DSP. The user interface shows memory, all the internal registers, I/O, etc. and the effect on these after each instruction is performed. The obvious disadvantage is that nothing happens in real time, so code cannot be tested as in the final application. Input/output operations are simulated using disk files, which can be awkward to set up and require a lot of interpretation.

The hardware platform can take the form of a so-called evaluation module or a true emulator. An evaluation module is a standard hardware platform that usually contains a processor, memory and analog I/O, but has limited possibilities for expansion. It may therefore not be able to truly mimic the final system. This is an appropriate tool for real-time testing of code before any prototype system hardware is available or if there is a strict limit on budget. Full real-time emulators are normally used when the software is to be tested on prototype target hardware, i.e., at the system integration stage. They allow the code to be tested in real time in its true system context, providing a high level of certainty of correct operation.

Writing and testing DSP code is a highly iterative process. With the use of a simulator or an evaluation board and a fairly powerful PC, code may be tested regularly as it is written. Writing code in modules or sections can help this process as each module can be tested individually, with a greater chance of the whole system working at the system integration stage. It is worth checking that the DSP you intend to use has an assembler and linker that fully support modular code, object libraries, etc.

It is also well worth looking at how user-friendly the simulator and evaluation boards are. There has been a trend for each different development tool for each processor to have a unique user interface. This leads to confusion for the engineer and a learning curve with any new development tool or processor. Texas Instruments has defined a common programmer's interface for all new development tools, which has even been retrofitted to some older tools. The programmer's interface is usually known as the C source debugger, which is an important element in code development and debug. All tools for all TMS320 processors other than the original TMS320C1x-generation tools have this common interface. New TMS320C1x tools will also have the interface, but without C language support since the TMS320C1x processors do not address sufficient memory space for C.

The debugger allows debugging at the C statement level, with simultaneous debugging of C and assembly language possible on one display screen. There is a clear display of the lines of assembler source related to each C statement, breakpoints may be set at the assembler or at the C level and so on. Code profiling tools can be used to display a histogram of time spent in each section of code, highlighting areas of C code which may need subsequent hand-optimization.

## HARDWARE DESIGN

It is not necessary to discuss here how to design the hardware of a DSP system. The basic requirements in terms of processor speed; memory size, speed and mix; analog I/O specification; and host processor support will have been worked out in the initial stages of system development. Implementing the application around a particular DSP is very device dependent. Some general hints are given in Chapter 7, but for detailed design information, refer to the user guide or data manual for your chosen device.

## SYSTEM INTEGRATION

This is the crunch time for any system. It is the first opportunity to try out the code with the application hardware. A full emulator is the best tool to use to test and debug the hardware and/or software, though many successful designs have been completed without an emulator.

Emulators take two basic forms. Those for DSP devices designed prior to about 1988 are traditional "in-circuit" emulators using an "umbilical" cord. These operate by replacing the processor in a target system, allowing full test and debug of code in the final system hardware. They can also work stand-alone for testing code that does not require any I/O access. They contain real-time hardware breakpoint and trace facilities, allowing a sophisticated level of code debug. It is possible to set a breakpoint or series of breakpoints based on a complex set, or sequence, of events. Code execution immediately prior to or immediately following these breakpoints may be traced. This provides a record of the sequence and timing of code execution. The key advantage is that this execution took place in real time so the trace is a true record of actual system behavior.

The real limitation of this type of emulator is a physical one. The umbilical cord does not work adequately for high-speed processors. Where clock frequencies approach 50 MHz the cable becomes a significant circuit element between the processor in the emulator and the memory and I/O in the target system by distorting interface timing and preventing correct operation.

Recently, a new approach to emulation has been adopted, which is sometimes referred to as in-system emulation. The emulator now simply acts as a controller for a processor device with on-chip emulation facilities and the device remains in the target system. Digital signal processors designed after about 1988 usually have these facilities built in. The emulator exerts its control through a serial scan path provided specifically for device test and emulation. Commands and data flow from the emulator to the device or attached memory through the scan path.

The processor has a series of internal registers allowing the emulator full access to, and control of, internal resources. By directly controlling the pins on the device, external memory may also be read, loaded or cleared. The processor can be single stepped and real-time breakpoints may be set. In short, all normal emulator facilities are provided, with one exception: Not all DSPs include real-time trace memory. Some processors have a small amount of trace memory on-chip, but as communications back to the emulator are non-real time, the trace length is limited to the size of this memory. In practice this is rarely a problem as improved breakpoint facilities, non-real time trace and single stepping can be used in combination to resolve any problems with the software.

The in-system serial emulation is an extension of the serial scan path system-testing protocol known as IEEE 1149.1 or JTAG after the Joint Test Action Group, which established the specification. Basically, the internal emulation circuits are accessed using the same serial port. The TMS320C5x and TMS320C4x devices from Texas Instruments are JTAG compatible and have the internal emulation path. TMS320C3x devices have just the emulator serial scan architecture known as MPSD, simply because the JTAG specification did not exist in 1987 when the TMS320C30 was designed.

## REFERENCES

There are very few general references in the area of DSP system design. Bateman and Yates [1988] makes a bold effort and contains much useful information. DeFatta et al. [1988] contains an excellent section on design methodology. Another useful approach is to look to general DSP training courses as provided by a number of independent training organizations.

For those wanting to use C, there is an excellent book by Embree and Kimble [1991] that contains a brief introduction to DSP algorithms and follows this up with a detailed description of how to implement these in C. It rather skips over the fact that most practical systems will be a mixture of C and assembler, but it is an essential book for anyone using C in a DSP application. Example source code in Kernighan and Ritchie C is provided on a disk.

There are also some independent books on designing with specific DSPs. The most substantial of these is Chassaing and Horning [1990], which deals with the TMS320C25 in great depth. This book was developed from a college lecture course and is a little mathematical in places. Despite this, it contains a wealth of practical information on using a fixed-point DSP device.

Manufacturers also publish extensive application volumes in support of their DSP devices. These are an extremely valuable source of information. At the time of this writing, there were four separate volumes available for TI DSPs.

Bateman, Andrew and Yates, Warren [1988], *Digital Signal Processing Design*, Pitman, London, UK.

Chassaing, Rulph and Horning, Darrell W. [1990], *Digital Signal Processing with the TMS320C25*, John Wiley, New York.

DeFatta, David J., Lucas; Joseph G. and Hodgkiss, William S. [1988]. *Digital Signal Processing: A System Design Approach*, John Wiley, New York.

Embree, Paul M. and Kimble, Bruce [1991], *C Language Algorithms for Digital Signal Processing*, Prentice-Hall, Englewood Cliffs, NJ.

# Glossary of Acronyms

| | |
|---|---|
| ACC | ACCumulator |
| ADC | Analog-to-Digital Converter |
| ADM | Adaptive Delta Modulation |
| ADPCM | Adaptive Differential Pulse Coded Modulation |
| AIC | Analog Interface Circuit |
| ALU | Arithmetic Logic Unit |
| AM | Amplitude Modulation |
| AQB | Adaptive Quantization – Backward |
| AQF | Adaptive Quantization – Forward |
| APC | Adaptive Predictive Coding |
| ASIC | Application Specific Integrated Circuit |
| ASK | Amplitude Shift Keying |
| ATM | Automatic Teller Machines or Asynchronous Transfer Mode |
| BPF | Band-Pass Filter |
| CCITT | Committee Consultative International Telegraph and Telecommunication |
| CD | Compact Disc |
| CDMA | Code Division Multiple Access |
| CELP | Code Excited Linear Predictive (coding) |
| CEPT | Conference on European Posts and Telecommunications |

| | |
|---|---|
| CISC | Complex Instruction Set Computer |
| CMOS | Complementary Metal Oxide Semiconductor |
| CODEC | COder DECoder |
| COMPAND | COMpress and exPAND |
| CPU | Core Processing Unit |
| CR | Capacitor/Resistor |
| CT2 | Cordless Telephone 2 |
| CVSD | Continuously Variable Slope Delta (modulation) |
| CW | Continuous Wave |
| | |
| DAC | Digital-to-Analog Converter |
| DCT | Discrete Cosine Transform |
| DECT | Digital European Cordless Telephone |
| DFDP | Digital Filter Design Package |
| DFT | Discrete Fourier Transform |
| DM | Delta Modulation |
| DMA | Direct Memory Access |
| DPCM | Differential Pulse Coded Modulation |
| DSP | Digital Signal Processor or Digital Signal Processing |
| | |
| ECL | Emitter Coupled Logic |
| ENIAC | Electronic Numerical Integrator and Calculator |
| EEPROM | Electrically Erasable Programmable Read Only Memory |
| EOB | End of Buffer |
| EPROM | Erasable Programmable Read Only Memory |
| | |
| FCT | Fast Cosine Transform |
| FD | Full Duplex |
| FDD | Frequency Division Duplex |
| FDM | Frequency Division Multiplexing |
| FDMA | Frequency Division Multiple Access |
| FDP | Fast Digital Processor |
| FIR | Finite Impulse Response (filter) |
| FFT | Fast Fourier Transform |
| FM | Frequency Modulation |
| FSK | Frequency Shift Keying |
| | |
| GMSK | Gaussian Minimum Shift Keying |
| GSM | Groupe Special Mobile or Global System for Mobile (communications) |

| | |
|---|---|
| HD | Half Duplex |
| HLL | High-Level Language |
| HPF | High-Pass Filter |
| | |
| IAS | Institute for Advanced Studies |
| IC | Integrated Circuit |
| IDCT | Inverse Discrete Cosine Transform |
| IDFT | Inverse Discrete Fourier Transform |
| IFFT | Inverse Fast Fourier Transform |
| IIR | Infinite Impulse Response (filter) |
| ISDN | Integrated Services Digital Network |
| ISO | International Standards Organization |
| | |
| JPEG | Joint Photographic Expert Group |
| JSRU | Joint Speech and Research Unit |
| JTAG | Joint Test Action Group |
| | |
| LAN | Local Area Network |
| LDM | Linear Delta Modulation |
| LMS | Least Mean Square |
| LP | Long Playing (record) |
| LPC | Linear Predictive Coding |
| LPF | Low-Pass Filter |
| LSB | Least Significant Bit |
| | |
| MAC | Multiply and ACcumulate or Multiplier/ACcumulator |
| MACD | Multiply ACcumulate and Data move |
| MBPS | Million Bits Per Second |
| MIPS | Million Instructions Per Second |
| MFLOPS | Million Floating-Point Operations Per Second |
| MLPC | Multipulse excited Linear Predictive Coding |
| MOPS | Million Operations Per Second |
| MPEG | Moving Pictures Expert Group |
| MPSD | Modular Parallel Scan Device |
| MPYK | Multiply Immediate with Constant |
| MSB | Most Significant Bit |
| | |
| NMOS | N-type Metal Oxide Semiconductor |
| NTSC | National Television Systems Committee |
| | |
| PABX | Private Automatic Branch eXchange |

| | |
|---|---|
| PAL | Phase Alternate Line |
| PAM | Pulse Amplitude Modulation |
| PC | Personal Computer |
| PCM | Pulse Coded Modulation |
| PCN | Personal Communication Networks |
| PDM | Pulse Density Modulation |
| PM | Pulse Modulation |
| PPM | Pulse Position Modulation |
| POS | Point of Sale |
| PSK | Phase Shift Keying |
| PSTN | Public Switched Telephone Network |
| PWM | Pulse Width Modulation |
| QAM | Quadrature Amplitude Modulation |
| QMF | Quadrature Mirror Filter |
| QPSK | Quadrature Phase Shift Keying |
| RAM | Random Access Memory |
| RELP | Residual Excited Linear Predictive (vocoder) |
| ROM | Read Only Memory |
| RPT | Repeat |
| RC | Resistor/Capacitor |
| RF | Radio Frequency |
| SBC | SubBand Coding |
| SDH | Synchonous Digital Hierarchy |
| SH | Sample and Hold |
| SONET | Synchronous Optical NETwork |
| SNR | Signal-to-Noise Ratio |
| SOB | Start Of Buffer |
| TDD | Time Division Duplex |
| TDM | Time Division Multiplexing |
| TDMA | Time Division Multiple Access |
| VLSI | Very Large Scale Integration |
| VSELP | Vector Sum Excited Linear Predictive (coding) |
| ZOH | Zero-Order-Hold |

# Index